Governing the Nile River Basin

Governing the Nile River Basin

The Search for a New Legal Regime

Mwangi S. Kimenyi *and*
John Mukum Mbaku

BROOKINGS INSTITUTION PRESS
Washington, D.C.

THE BROOKINGS INSTITUTION
1775 Massachusetts Avenue, N.W., Washington, D.C. 20036
www.brookings.edu

Library of Congress Cataloging-in-Publication data is available.

ISBN: 978-0-8157-2655-5 (pbk.: alk. paper)

9 8 7 6 5 4 3 2 1

Printed on acid-free paper

Typeset in Minion

Composition by Oakland Street Publishing
Arlington, Va.

To the peoples of the Nile River Basin

with the hope that they may find the courage and foresight to come together
and negotiate in good faith to develop and adopt a legal and institutional framework
that will enhance their ability to live together peacefully,
allocate the waters of the Nile River Basin in a reasonable, equitable,
and sustainable manner, and
create the wealth needed to deal effectively and fully
with poverty and deprivation.

Contents

Preface and Acknowledgments

This monograph deals with an issue that has become increasingly important both for Africa's economic development and for the peaceful coexistence of millions of Africans who reside in the Nile River basin. At the heart of the issue is the efficient and equitable allocation of the waters of the Nile River among the various riparian states. This issue is currently contentious because the basin's legal regime consists of treaties and agreements that are not recognized by all the riparian states. These treaties, collectively referred to as the Nile Waters agreements, grant rights to the utilization of the waters of the Nile River exclusively to Egypt and the Republic of Sudan and hence severely disadvantage other riparian states, a state of affairs that objective observers and analysts agree is not tenable. Given this fact, an alternative framework for the allocation of the Nile River waters is called for. The primary goal of this study is to propose an approach to reaching such a legal framework that is acceptable to all the riparians.

But reaching an agreement on the Nile River waters will be no small feat. Any changes to the allocation of the waters of the Nile River must necessarily involve trade-offs—with beneficiaries and losers. Owing to climate change, the amount of water available in the future is likely to be lower than the current volumes, and thus all riparians will need to make adjustments. Because achieving a Pareto allocation—one that makes one country better off without making another worse off—is not feasible, the goal would be to agree on frameworks that minimize losses and maximize collective benefits to the riparians.

Given the current system of allocation, in which virtually all of the waters of the Nile River are allocated to Egypt and the Republic of Sudan, an opportunity exists for reallocations that would generally increase collective welfare without causing substantial damage to the two countries. From a utilitarian perspective, reducing the amount of water available to Egypt and Sudan by only a small amount and allocating the same to countries that currently do not have access to the water would be associated with overall gains in collective welfare.[1] Basically, the losses to the two downstream countries would be more than outweighed by the gains to the upstream riparian states. In essence, there is a justifiable normative basis to engage in negotiations to create a new legal framework.

Emotions tend to run high when it comes to the waters of the Nile River. On one hand, Egypt and the Republic of Sudan insist that allocation be governed by the Nile Waters agreements, and on the other hand, the upstream riparian states argue that they are not bound by those agreements and that all riparian states should engage in inclusive negotiations to create a new agreement, one that they believe would allow for more equitable and reasonable allocation and utilization of the waters of the Nile River. The dynamics in today's Nile River basin point to an urgent need for a negotiated alternative legal framework for water allocation and use. Continued insistence by Egypt and the Republic of Sudan that the upstream riparian states be severely limited in the way they harvest and utilize the Nile River waters that originate in, or pass through, their territories is not justifiable on either moral or legal grounds. Despite threats of military action by Egypt, Ethiopia has decided to proceed with the construction, on the Blue Nile, of the Grand Ethiopian Renaissance Dam, which is expected to be completed in 2017. Other Nile River riparian states are likely to want to engage in similar projects. The most effective way forward for the basin calls for all countries to work together to design new water-sharing arrangements to ensure equity, fairness, and sustainability.

Although we are critical of the colonial-era Nile Waters agreements—which we do not consider to be relevant to most of the riparians—we emphasize that our proposed solution is not to overhaul the water-sharing agreements in a manner that disrupts or severely penalizes the livelihoods of the people of Egypt and Sudan. The importance of the Nile River, especially to Egypt, is well recognized, and thus we advocate for a consultative process that protects current beneficiaries while simultaneously increasing the amount of water available to others. Thus though we have at times used relatively strong language to demonstrate the position taken by Egypt as regards proposals to change the legal framework for allocating the waters of the Nile River, this should not be taken to mean that we are anti-Egypt or that we

1. The basic premise of this argument is the well-known economic principle of diminishing marginal utility.

oppose the interests of the Egyptian people. However, we emphasize that Egypt must change its position to accommodate the views of other riparians and also to appreciate their development needs.

In preparing this monograph, we have received helpful comments and suggestions from colleagues and external reviewers. We wish to thank John MacArthur of the Brookings Institution and Robert Adler, dean of the S. J. Quinney College of Law, University of Utah, and Marc A. Levy, Center for International Earth Science Information Network, Columbia University, for insightful and useful comments and suggestions. We have attempted to address many of the suggestions that the reviewers provided. This process has significantly improved the study by clarifying a number of issues.

However, owing to the intended specific focus of the study, there were many important suggestions that we did not incorporate in the study. Reviewers, for example, proposed that we should provide detailed discussion of the hydrology of the Nile, the historical uses of the Nile waters, and more in-depth discussion about other transboundary water agreements. For this study, we have opted to focus primarily on the legal framework governing the allocation of waters of the Nile River, which is the contentious issue that we believe demands urgent attention. Many of the various issues raised by the reviewers have already been covered by experts in the respective areas.

Reviewers also expressed concern about the negative or what may appear as a biased way Egypt is portrayed in the study. In particular, some reviewers expressed concern about statements that portray Egypt as adamant, unreasonable, war-mongering and unwilling to accommodate the views of other riparians on the matter of allocation of the waters of the Nile River. Our position is that it is important to provide an accurate picture of the views of Egypt and other riparians. As much as is possible, statements about the position taken by various riparians is supported by relevant sources. We believe that to effectively deal with the challenge of arriving at a new legal framework for governing the Nile River basin, it is important to be explicit and categorical about the attitudes and views of the various riparians. This is the way we have presented the position taken by Egypt and other riparians.

The matter of the allocation of the waters of the Nile River is a serious one, and failure to arrive at new water allocation arrangements could easily trigger armed conflict. Furthermore, the failure of the Nile River basin countries to reach consensus on how to allocate the river's waters could result in inefficient and wasteful use of the common pool resource, making all worse off.

This monograph reflects only the present views and considerations of the authors, which should not be attributed to either Weber State University or the Brookings Institution or any other institution with which they are associated.

1 *The Political Economy of Transboundary Water Resource Management in Africa*

Most African countries have territory that is located in at least one transboundary water basin (see, for example, Lautze and Giordano 2005, p. 1053), and about 62 percent of the continent's land mass is covered by transboundary water basins (Wolf and others 1999, p. 392). Because of the pervasiveness of transboundary water basins in the continent, "African water management is also, by definition, transboundary water management" (Lautze and Giordano 2005, p. 1054). Hence most water law on the continent has, historically, been transboundary water law.

One can begin the study of water management in Africa by taking a look at existing legal regimes that regulate the allocation of water across the continent. Ordinarily, that would lead one to those institutional arrangements that deal with transboundary water basins—that is, transboundary water law. African transboundary water law consists of agreements and treaties that were concluded in both colonial and postcolonial periods, many international water law conventions and treaties, and various customs and traditions that have, throughout history, regulated water use. To date, researchers have identified "more than 150 agreements, treaties, protocols, and amendments spanning over 140 years and involving more than 20 African basins" (Lautze and Giordano 2005, p. 1054).

During the colonial period, European countries that undertook the development and implementation of transboundary water agreements often did so not to ensure the fair allocation of water to benefit the African populations but to maximize European objectives in the colonies. As a consequence, the settling of boundaries between territories claimed by one European colonizer or another often dominated some of these agreements. For example, the

1891 Anglo-Italian protocol (officially referred to as the Protocol between the Governments of Great Britain and Italy, for the Demarcation of Their Respective Spheres of Influence in East Africa, from Ras Kasar to the Blue Nile) was designed by the two colonial powers not only to deal with water issues but also to settle the boundary between Italian Eritrea and British Sudan. The 1902 Anglo-Ethiopian treaty (Treaties between the United Kingdom and Ethiopia and between the United Kingdom, Italy, and Ethiopia relative to the Frontiers between the Soudan, Ethiopia, and Eritrea) also deals with boundary determination: the treaty aimed to settle the boundary between Sudan, which was at the time a British colony, and Ethiopia.

Jonathan Lautze and Mark Giordano (2005, pp. 1075–87) provide a relatively comprehensive list of African transboundary water agreements and treaties. Our interest in this monograph is not to delve into all the water agreements or into all the continent's water basins. Instead, we take a look at the agreements surrounding the Nile River basin that have regulated the allocation of its waters. Specifically, we provide an overview of the Nile Waters agreements—the 1929 Anglo-Egyptian treaty and the 1959 bilateral agreement between Egypt and Sudan, which the two countries claim to be the main legal framework for the Nile River basin.[1] Today's Nile River riparians, except for Egypt and the Republic of Sudan, consider these agreements anachronistic holdovers from the colonial era and want them abrogated and replaced by a new international watercourse legal regime that enhances equity in the allocation of the Nile River's waters. Egypt and Sudan, however, insist that the existing Nile Waters agreements be maintained or that, in the event a new legal regime is established, Egypt's historical rights—those granted by the original agreements—should be honored.[2]

Although the Nile Waters agreements specifically mention Egypt's "acquired rights," that virtually all upstream riparian states have renounced these agreements and do not consider them binding brings into question the

1. Officially, Exchange of Notes between His Majesty's Government in the United Kingdom and the Egyptian Government in Regard to the Use of the Waters of the Nile River for Irrigation Purposes (with Seven Diagrams), May 7, 1929, L.N.T.S. 2103; and United Arab Republic and Sudan Agreement (with Annexes) for the Full Utilization of the Nile Waters, Cairo, November 8, 1959, 6519 U.N.T.S. 63. The short forms are used throughout this volume.

2. Although treaties grant rights, those rights are granted as between those states that are parties to the treaties and/or are bound by them. Since the upstream riparian states have made it clear that, as nonparties to these agreements, they are not bound by them, the validity of these rights is in question. Hence, they do not recognize the "historical rights" claimed by both Egypt and the Republic of Sudan.

validity of Egypt's claims.[3] All property rights are relative, and treaties or other agreements may grant rights only as between those states that are actually bound by the treaties granting such rights and cannot do so relative to parties that are not bound by the treaties.

The Nile River basin's existing legal arrangements do not provide the wherewithal for the effective management of the basin's multifarious problems, which include the allocation of water, climate change, ecosystem degradation, and resource sustainability. We provide guidelines for the construction of an effective and viable legal mechanism that is capable of achieving fairness and sustainability in the allocation and utilization of the waters of the Nile River, as well as meeting the needs of the basin's economies, which are searching for ways to improve the living standards of their citizens. Such an agreement, we believe, would be acceptable to all riparian states.

Here we examine the failure of the countries of the Nile River basin to provide a legal regime that is acceptable to all and provides the necessary mechanisms for the equitable, fair, reasonable, and sustainable utilization of the waters of the Nile River.[4] In chapter 2, we provide an overview of the physical characteristics of the Nile River, its tributaries, and sources. We also provide information on the river's riparian states and briefly examine various activities, such as agriculture, that affect water use in the basin. Finally, we examine the impact of climate change on the basin, generally, and water use, in particular.

In chapter 3, we explore various historical events that have contributed to the nature of conflict, specifically that related to the use of water, in the Nile River basin. For example, we take a look at how the U.S. Civil War created opportunities for the development of cotton production in the Nile River basin, significantly changed the political economy in the region, and set the stage for the conflict that currently afflicts the basin.

In chapter 4, we examine the Nile Waters agreements, which are considered a key to understanding the basin's present conflict. Although the downstream riparian states argue that these bilateral treaties represent the basin's legal regime, the upstream states reject that claim, argue that they are not bound by them, and seek to produce a new inclusive legal framework.

Chapter 5 is devoted to an examination of theories of treaty succession and their possible impact on governance in the Nile River basin. Of special

3. See chap. 1, para. 2, of the 1959 bilateral agreement between Egypt and Sudan.

4. A major problem here is that it is not possible to maintain current allocations—that is, those provided by the Nile Waters agreements—and still increase access to the waters of the Nile River for the upstream riparian states. Equity and fairness necessarily imply trade-offs, which must involve a certain level of sacrifice by both Egypt and the Republic of Sudan.

interest is the Nyerere doctrine and how it was used by Britain's former colonies to justify their rejection of treaties that were entered into on their behalf by Britain.

In chapter 6, we examine international water law and its implications for governance in the Nile River basin. Specifically, we examine the UN Convention on the Law of Nonnavigational Uses of International Watercourses and determine the types of insights that it can provide for the Nile River basin countries as they struggle to develop an inclusive legal framework for the basin.[5]

In 1999 the Nile Basin Initiative was signed by the Nile River basin riparian states (except Eritrea) as a mechanism to enhance the equitable, fair, and sustainable utilization of the waters of the Nile River. In chapter 7, we examine the initiative and its relevance to the effective resolution of water-related issues in the Nile River basin.

Chapter 8 is devoted to an examination of the Cooperative Framework Agreement (CFA). In taking a look at this new agreement, we try to resolve the question of whether it can serve as the inclusive legal instrument that would finally bring an end to the struggle between the upstream and downstream states over how to allocate the waters of the Nile River.

In chapter 9, we review the tumultuous relationship between Egypt and Ethiopia, which over the years has had a significant impact on water-related conflicts in the basin. Chapter 10 is devoted to an examination of the Grand Ethiopian Renaissance Dam, a project that is currently under way and is expected to have a significant impact on the demand for and supply of water in the Nile River basin.

Finally, in chapter 11, we suggest a way forward for the Nile River riparian states. Specifically, we conclude from our study that all relevant stakeholders—the upstream and downstream riparians—should engage in negotiations to produce a new inclusive treaty that would provide them with an effective legal mechanism for regulating the use of the waters of the Nile River.

Our goal is not merely to be critical of the current legal regime—the Nile Waters agreements—or to advocate the development and implementation of legal frameworks that would jeopardize the livelihoods of the people of Egypt and the Republic of Sudan or any other riparian state. Instead, we seek to show that inclusive negotiations would produce a legal regime capable of providing an environment leading to equitable, fair, and reasonable management of the Nile River waters and the peaceful coexistence of the populations of the states that share this common resource.

5. UN Convention on the Law of the Nonnavigational Uses of International Watercourses, New York, May 21, 1997, G.A. Res. 51/229, U.N. Doc. A/RES/51/229.

2 Physical Description of the Watercourse and Basin States

The Nile River basin consists of complex ecosystems that are characterized by "high climatic diversity and variability, a low percentage of rainfall reaching the main river, and an uneven distribution of its water resources" (NBI 2012, p. 26). It is home to thousands of plant and animal species, many of which are endemic to the basin.

The Nile River basin's environmental resources and its water provide the basin's diverse peoples with a rich variety of goods and services and contribute significantly to the region's gross domestic product. The basin's system of waterways and wetlands provides both a flight path and a destination for migratory birds from other parts of the continent.

For many years, the Nile River basin's natural resource base has been threatened, and continues to be threatened, by various pressures, including agriculture, population increases, urbanization, invasive species, bushfires, mining and quarrying, climate change, and natural disasters. Although several game, wildlife, and forest reserves, as well as national parks, have been established to protect and conserve the Nile River basin's resources and its unique ecosystems, degradation remains a major problem for the basin (NBI 2012). Rapid degradation of the basin's ecosystems has become an important concern for many of the basin's states. Unfortunately, population pressures, relatively weak governments, dysfunctional legal and institutional frameworks, and poor public policies in the riparian states continue to constrain and endanger ecosystem management in the basin.

The irrigation of agricultural fields in Egypt and the Republic of Sudan represents the single most important use of the waters of the Nile River—

primarily because these two riparian states have the most developed irrigation systems in the basin.[1] In recent years, however, many upstream riparian states have either developed or are in the process of developing the capacity to more effectively harvest and utilize the Nile River's waters for national development.[2] As the upstream riparians develop the capacity to challenge the downstream states for the allocation of the river's "renewable discharge" (NBI 2012, p. 26), issues of equitable, fair, and reasonable use of the waters of the Nile River, as well as the sustainability of the basin's ecosystem, are likely to intensify.

The Nile River and Its Basin

The Nile River is the world's longest watercourse, measuring 4,160 miles (6,695 kilometers) and flowing over thirty-five degrees of latitude (Collins 2002, p. 11; see also figure 2-1). Its basin covers an area of about 1.227 million square miles (3.18 million square kilometers), approximately 10 percent of the African continent, and is shared by eleven countries of the river's basin.[3] The river's southernmost source is a spring in Burundi called Kikizi (see figure 2-1). The Kikizi spring and others like it eventually coalesce into rivers. Two of these rivers, the White Nile, whose ultimate sources are in Burundi, and the Blue Nile, which originates in the Ethiopian highlands, are the main sources of the Nile River's waters. The White Nile, which flows through equatorial Africa, also runs through Lakes Albert and Victoria, the latter of which is shared by Kenya, Tanzania, and Uganda. The White Nile provides about 15 percent of the waters that flow into the Nile River, and the Blue Nile, whose

1. It is also possible that the ability of Egypt and the Republic of Sudan to develop their irrigation infrastructure was made possible by the Nile Waters agreements, which have favored them in water allocation. The development of irrigation systems in Egypt and Sudan began during the colonial period, shortly after the start of the Civil War in the United States, when London took an interest in the development of agriculture in the region, specifically, the potential of cotton growing to replace what had been imported from the American South.

2. The upstream riparian states are Burundi, Democratic Republic of the Congo, Eritrea, Ethiopia, Kenya, Rwanda, Tanzania, and Uganda. The downstream riparian states are Egypt and the Republic of Sudan. In 2011 several of Sudan's southern provinces, which had been fighting with Khartoum about issues of self-determination, were granted independence and took the name South Sudan. Although the new country is a downstream riparian, it has tended to side with the upstream riparians regarding Nile River governance issues.

3. These countries are Burundi, Democratic Republic of the Congo, Egypt, Eritrea, Ethiopia, Kenya, Rwanda, South Sudan, Sudan (Republic of), Tanzania, and Uganda.

Figure 2-1. *The Nile River Basin*

tributaries include the Abbay, Sobat, and Atbara Rivers, provides 85 percent of the water flowing into the Nile River as measured at Aswan in Egypt (NBI 2012, p. 36).[4]

The Nile River basin consists of "distinct regions that correspond to the different typography through which it passes on its long journey out of equatorial Africa to the Mediterranean Sea" (Collins 2002, p. 1; see also NBI 2012; and figure 2-1). In its southernmost source, the Nile River starts as many springs and rivers, "plunges down as the Bahr al-Jabal, the Mountain River, into the swamps of the southern Sudan known as the Sudd, from which it emerges into the broad expanse of the White Nile and its great meeting at Khartoum with the Blue Nile from Ethiopia" (Collins 2002, p. 1). From Khartoum, the Nile River flows to the Nile delta and into the Mediterranean Sea (Sutcliffe and Parks 1999, p. 1; see also figure 2-1).

From Burundi and Rwanda, the many rivers and springs that form various tributaries of the Nile River converge into the Kagera River, which flows into Lake Victoria, the world's second-largest freshwater lake (Collins 2002, p. 2). Within this plateau are other lakes, including Lakes George, Edward, Albert, and Kyoga. Although these lakes contribute significantly large amounts of water, the Nile River flows directly from the northern tip of Lake Victoria as the Victoria Nile. The latter "pours over the Owen Falls Dam and then into Lake Kyoga, after which, in a spectacular drop, it crashes down from the lip of the great African Rift Valley at Kabarega Falls (formerly Murchison Falls) to its bottom at Lake Albert," and strengthened by waters from several lakes, notably, Lakes George, Edward, and Albert, "the Victoria Nile flows out of the Lake Plateau as the Albert Nile to the border of Uganda at Nimule, where it becomes the Bahr al-Jabal, or the Mountain River" (Collins 2002, p. 2; see also figure 2-1).

The Mountain River struggles slowly through hundreds of miles of swampland and eventually flows into Lake No, a relatively shallow water body, whose main outlet is the White Nile. The latter flows toward the Mediterranean Sea, and some seventy-five miles from its start at Lake No, its flow is strengthened significantly by waters from the Sobat River, which originates in Ethiopia. Now totally replenished by waters from the Sobat, the White Nile flows north for nearly five hundred miles through South Sudan and the Republic of Sudan and eventually joins the Blue Nile at Khartoum and becomes the Nile River (see figure 2-1).

The Blue Nile originates in the Ethiopian highlands at about six thousand feet. The Blue Nile, called the Great Abbai in Ethiopia, flows out of the south-

4. At Aswan, the annual average flow is 84 billion cubic meters of water.

ern shore of Lake Tana, runs in a southeastern direction, drops over the Tisisat Falls, and eventually curves until it begins to flow north toward Khartoum, and at Khartoum, the two rivers unite to form the Nile River. About two hundred miles north of the confluence of the two rivers at Khartoum, the Nile's last major tributary, Ethiopia's Atbara River, pours its waters into the Nile River, notably after the rains from the Semyen Massif in Ethiopia. Below the Atbara, the Nile takes an S-shaped turn and flows "grandly through the Nubian and Egyptian deserts, where no rain or other rivers provide relief from the desiccation" (Collins 2002, p. 3). This part of Africa consists of a hostile and unforgiving desert, but the Nile River has provided the wherewithal to grow impressive civilizations, including those cultivated by the pharaohs, the Greeks, Romans, and Arabs, all of which relied on the Nile River for survival (Collins 2002, p. 3).

The Nile has a complex hydrology, which is characterized by an annual flood. The Nile basin has only one rainy season, which usually occurs in the summer, after heavy rains in the Ethiopian highlands have filled many rivers, including the Sobat, the Blue Nile, and the Atbara. The waters in these rivers make their way into the Nile River, which begins to rise in May in the northern part of the Republic of Sudan and in June in southern Egypt at Aswan (Collins 2002, p. 4). The Nile River reaches its maximum water level in August and September and begins to decrease thereafter. Even at the height of the rainy season, the volume and reliability of precipitation declines as one moves northward from the Ethiopian highlands and the western forests of South Sudan to the relatively arid regions of Egypt and the northern parts of the Republic of Sudan, where there is little rainfall (NBI 2012). From the Ethiopian highlands westward to the forests of South Sudan, and northward to the Republic of Sudan and on to Egypt, the "pattern of vegetation and distribution of surface water bodies" clearly illustrates the "spatial variability of rainfall" in the Nile River basin (NBI 2012, p. 30).

Although the occurrence of the flood and its timing have been predicted with relative accuracy, it has been quite difficult to determine the volume of water that would be delivered by each flood. During good years, just enough water can flow into the Nile River from the Ethiopian highlands to cause the river to overflow its banks and create within the surrounding basin land that enhances abundant crop production. However, in some years, the amount of water rushing from the Ethiopian highlands into the Nile River is so much that it destroys homes and other properties and causes misery in the communities along the river (Collins 2002; Sutcliffe and Parks 1999; Melesse 2011; Awulachew and others 2012).

The completion of the Aswan High Dam in 1970 created Lake Nasser, an enormous reservoir, which stretches from southern Egypt into northern Sudan, where the reservoir is called Lake Nubia (see figure 2-1). The dam has protected against some of the basin's most destructive floods and has also provided the region with hydroelectric power and water for year-round irrigated agriculture.

Despite the enormous benefits that it has provided to Egypt, the Aswan High Dam has imposed significant costs on communities in both Egypt and Sudan; its construction resulted in the forced relocation of Sudanese Nubians. In addition, the dam has produced many negative externalities, including "decreased fertility in the surrounding agricultural lands, serious erosion downstream, inundation of the delta with seawater, increases in soil salinity and water-logging, increased seismic activity from the weight of the reservoir on fault lines, and decreases in fish populations in the Mediterranean Sea" (Collins 2002, p. 4).

From the Aswan High Dam, the Nile River flows due north for about five hundred miles. The yearly overflow of the river's banks has created a relatively narrow floodplain, full of rich and productive soil. Beyond both sides of this floodplain is desert, which is not suitable for agriculture and human habitation. The river runs straight into Cairo, the center of Egyptian civilization. The Nile delta, located just north of Cairo, is "an enormous triangular region of flat alluvial plain, the most fertile soil in Africa, more than fifty feet deep and built up over thousands of years by the flooding river and its deposits of nutrient-laden silt from upstream" (Collins 2002, p. 5). This delta has supported Egyptian civilizations for thousands of years.

The Population of the Basin States

The population of the Nile River riparian states is 437 million, which represents about 41 percent of the African population (NBI 2012, chap. 4). Of the eleven countries that form the Nile River basin, Ethiopia has the highest population (86.5 million) and Egypt, the next highest, at 83.9 million. At the other end, Burundi (8.7 million) and Eritrea (5.6 million) have the smallest populations (see table 2-1).

Of a total population of 437 million, 238 million (or 54 percent) live within the basin area, and the percentage of the population of each riparian state that resides within the basin ranges from a high of 99 percent for Uganda to a low of 4 percent for the Democratic Republic of the Congo. Egypt has the highest number of people who live in the basin (80.4 million); Uganda has

Table 2-1. *Population of the Nile River Basin as Share of Total Population, by Country, 2012*

Country	*Population*	*Living in Nile River basin (percent)*	*Living in rural areas (percent)*
Burundi	8,749,387	58.8	89
DRC	69,575,394	3.8	66
Egypt	83,958,369	95.7	57
Eritrea	5,580,862	37.6	79
Ethiopia	86,538,534	40.3	83
Kenya	42,749,418	39.7	76
Rwanda	11,271,786	82.6	81
South Sudan	9,614,498	99.0	82
Republic of Sudan	36,107,585	87.3	67
Tanzania	47,656,367	21.5	73
Uganda	35,620,977	99.4	84

Source: Nile Basin Initiative (2012, p. 240). Note that until 2011, South Sudan was part of the Republic of Sudan. Between January 9 and 15, 2011, the people of South Sudan voted overwhelmingly in a UN-supervised referendum to secede from Sudan and gain their independence as a sovereign, independent state. South Sudan officially became an independent state on July 9, 2011.

35.4 million, Ethiopia 34.6 million, and the Democratic Republic of the Congo, 2.6 million (NBI 2012, p. 240).

Of the 238 million people who live in the Nile River basin, 172 million or 72 percent, reside in rural areas. Among all the riparian states, Burundi has the highest proportion of rural citizens (89 percent), and Egypt has the lowest (57 percent) (Collins 2012, p. 240). Although rural-to-urban migrations are rising rapidly in the basin, the region is expected to remain essentially rural for many years to come. Of course, there are exceptions: both the Republic of Sudan and Egypt are expected to see significant drops in their rural populations. Rural populations in Burundi, Ethiopia, Kenya, Rwanda, and Uganda are expected to dominate even by 2050 (NBI 2012, pp. 104–05).

The Nile River basin consists primarily of relatively underdeveloped and poor countries. There is, however, significant variability in levels of human and economic development (see, for example, UNDP 2013, pp. 144–47). According to the United Nations Development Program's human development index, only Egypt, among the Nile River riparian states, falls in the "medium" group, and it is one of the few countries among the riparian states that has been able to make substantial improvements in the quality of life of its citizens since the mid-1980s. The other riparian states fall within the "low human development" group, with eight of these countries ranked among the world's twenty-five poorest and least developed countries.[5] Most of Egypt's population lives on the narrow tract of the fertile land that straddles both sides of the Nile River and the Nile delta, and the country's economy also benefits significantly from a rising oil sector. Meanwhile, Tanzania, Kenya, Uganda, and Ethiopia, headwaters countries, have not been able to significantly improve living conditions for their citizens, most of whom live in communities that are scattered across difficult and hard-to-reach terrain (NBI 2012; UNDP 2013).

Most of the Nile River riparian countries are relatively poor. In 2012 the per capita gross national income (GNI) for the subregion called sub-Saharan Africa was U.S. $2,010.[6] Of all the Nile riparian states, only Egypt ($5,401) was able to meet and exceed the average for sub-Saharan Africa (UNDP 2013, pp. 145–46). That the basin has some of the continent's poorest economies is reflected in its relatively low per capita GNI and human-development-index values, which place all but one of its countries in the low human-development group. A significant proportion of the basin's population lives on less than U.S. $1.25 a day; in 2012, that population ranged from a high of 81 percent in Burundi (GNI per capita of $544), 77 percent in Rwanda (GNI per capita of $1,147), to a low of 2 percent in Egypt (GNI per capita of $5,401) (UNDP 2013, pp. 145–46; NBI 2012; see also NBI, chap. 4). Most people in these countries depend on the rural agricultural sector for their living. Unfortunately, public policies since these countries gained independence have neglected investment in the agricultural sector, and as a result, poverty rates in the rural areas have increased significantly. Many of these countries, like their counterparts elsewhere on the continent, have exploited their rural agricultural sectors for the benefit of those who live in the urban sectors (see, for example, Bates 1981, 1983).

Agriculture and Water Use in the Nile River Basin

Agriculture remains the mainstay of virtually all of the states of the Nile River basin: this sector, located almost exclusively in the rural areas, provides employment for more than 75 percent of the total labor force of all the Nile River riparian states (NBI 2012, p. 109). Within the Nile River basin, Egypt,

5. UNDP (2013, pp. 145–47). The medium rank comprises countries that have achieved a human development index falling between 0.536 and 0.710. States ranked low are those with a human development index of less than 0.536.

6. Per capita GNI in this paragraph is given in 2005 purchasing power parity U.S. dollars.

which has a relatively developed agricultural sector, is the only economy that has been able to achieve a level of economic diversification that has lessened its dependence on agriculture. Recent data show that only 32 percent of Egypt's total labor force is employed in agriculture; compare that to Kenya and the Republic of Sudan, where 75 and 80 percent, respectively, of the economically active population are engaged in agriculture (NBI 2012, p. 109).

The single biggest and most important consumer of water in the Nile River basin is agriculture. This sector provides the inputs for the important agroindustrial sector, employs a large proportion of rural households, provides food for each riparian state's citizens, produces a significant share of nonfarm jobs (for example, the transportation of farm products to the urban areas for sale), and contributes to and strengthens regional integration through trade (NBI 2012, p. 123).

The withdrawal of water from the Aswan dam to irrigate agriculture accounts for as much as 78 percent of the peak flow of the Nile River (NBI 2012, p. 123). As the population of the Nile River basin countries has increased, especially in Egypt, which has the most sophisticated irrigation systems, there has been an accompanying increase in the demand for water, not only for food production but also for household use. Within the next several decades, it is expected that the demand for water from the Nile River will intensify as the demand for more food to meet the needs of a rising and bourgeoning population increases. Additional demands are likely to be placed on the Nile River's waters for nondomestic and nonagricultural uses, especially for water to feed rising industrial production throughout the basin.

Within the Nile River basin, most irrigated agriculture is undertaken in Egypt and the Republic of Sudan, primarily in the Nile delta (Egypt) and Gezira, in the Republic of Sudan (NBI 2012, pp. 126–28). Crops grown on these irrigated fields include wheat, fodder, maize, cotton, rice, vegetables, sorghum, groundnuts (peanuts), fruits, pulses, citrus, and sugarcane. In terms of yields, Egyptian irrigated agriculture remains the most productive in the Nile River basin; in 2012 irrigated farms in Egypt achieved a productivity of 7.6 tons per hectare, compared with 1.0–2.1 tons per hectare in Gezira (NBI 2012, p. 128).

The demands placed on the Nile River basin by agriculture must be viewed from two critical perspectives: First, water harvest and utilization in agriculture will generate negative externalities, which are likely to include surface and groundwater pollution, destruction of the topsoil, and problems of salinity, and these issues must be dealt with fully to enhance sustainability in the management of the basin's water resources. Second, agriculture will, in the foreseeable future, continue to exert a significant demand on the basin's freshwater resources (NBI 2012).

Agricultural systems in the Nile River basin range from mixed smallholder subsistence rain-fed systems to medium-to-large-scale smallholder and commercial irrigated farming systems. Rain-fed farming systems are the most popular types of agriculture practiced in the Nile River basin. More than 87 percent of the land cultivated in the basin falls under rain-fed agriculture and provides the livelihood for most of the rural inhabitants of the upstream riparian states (NBI 2012, p. 125). Within the Nile River basin, one can find the following rain-fed agricultural systems:

—mixed smallholder subsistence rain-fed systems, which are found primarily in the "sub-humid and humid parts of the Nile basin at altitudes between 500 and 1,500 meters above sea level" (NBI 2012, p. 125), on which farmers produce primarily cereals and legumes, as well as livestock, mostly for household consumption

—mixed highland smallholder subsistence rain-fed systems, found primarily in the highlands of Ethiopia and Eritrea, as well as the Equatorial Lakes region, at above 1,500 meters altitude

—forest-based farming systems, which are found primarily in southwestern Ethiopia and are dependent almost exclusively on the exploitation of forest ecosystems for various crops and animals

—mechanized rain-fed systems, found mainly in the eastern and western regions of the Republic of Sudan and South Sudan and dominated by crops such as coffee, tea, oil palm, rubber, and a few cereals and fruits, which are produced primarily for export

—irrigated farming systems, which are the largest users of water resources of the Nile River basin.

It is estimated that there are about 4.9 million hectares of land under irrigation in the Nile River basin, with an additional 0.7 million hectares that are equipped for irrigation but have not yet been irrigated. That brings the total amount of irrigable land in the basin to 5.6 million hectares (NBI 2012, p. 126). Most of the irrigated land is located in the Republic of Sudan (1,749,300 hectares) and Egypt (2,963,581 hectares). The only other Nile riparian states with significant amounts of irrigated land are Ethiopia (90,769 hectares), Kenya (34,156 hectares), and Uganda (25,131 hectares). Overall, Egypt and the Republic of Sudan account for most of the irrigated land in the basin (97 percent); the upstream riparian states account for only 3 percent of the basin's irrigated land (NBI 2012, p. 126).

For many years, the countries in the Equatorial Lakes Plateau (countries in the headwaters region of the basin—the upstream riparians) had a relatively reliable supply of rainwater for farming and hence did not invest in the development of irrigation infrastructure. Owing to the effects of climate change and other factors, many of the upstream riparian states have begun to develop agricultural policies that call for investment in irrigation systems and irrigated agriculture.

South Sudan, the basin's and the world's newest independent country, has the potential to develop into an important center for food production in the East Africa region. South Sudan, which currently receives most of its public revenues from oil, hopes to develop its agricultural sector, reduce the economy's reliance on oil, and evolve into the region's breadbasket (NBI 2012, p. 127).[7]

South Sudan has at least five ecological zones: rainforest, savannah forest, flood plains, wetlands, and semidesert. Roughly 90 percent of South Sudan's land area is generally considered suitable for agricultural activities. Nevertheless, only 4 percent of the country's land area is currently under cultivation. Most of the cultivable land stretches along its flood plains and is ideally suitable for both rain-fed and irrigated agriculture, and soil and climate conditions allow for the production of a wide variety of both food and cash crops—tobacco, vegetables, maize, cowpeas, rice, sugarcane, and banana. However, productivity on farms remains low. Rainfall is erratic, and irrigated farming is still in its embryonic stages of development and currently accounts for only 3 percent of the total area under cultivation (NBI 2012, p. 127).

South Sudan, like other Nile River riparian states, has recognized the importance of irrigation in improving agricultural production and enhancing food security, as well as generally providing the wherewithal for the country to deal fully and effectively with poverty and deprivation. Like its fellow Nile River riparian states, South Sudan is in the process of developing an irrigation master plan, which is expected to increase the demand for water from the Nile River and its tributaries and challenge Egypt's and the Republic of Sudan's hegemonic control of the river.

The Nile River basin faces any number of constraints to agricultural productivity. Among these are high dependency on rain-fed agriculture,

7. Until South Sudan shut down oil production in January 2012 over its continuing conflict with the Republic of Sudan over the use of the latter's pipelines for transmission of oil to export markets, oil revenues have represented 82 percent of South Sudan's GDP and 98 percent of government revenues. See "South Sudan: Extractives in Ten Minutes, April 2013," Open Oil (http://openoil.net/wp/wp-content/uploads/2013/07/South_Sudan_Extractives_in_Ten_Minutes.pdf).

widespread watershed degradation, low soil fertility, prevalence of pests and disease, small land holdings, irregular irrigation water supply, unpredictable prices, poor physical infrastructure, insecure land tenure, poor or nonexistent access to credit, high cost and poor quality of inputs, weak and limited agricultural extension services, and insecurity arising from civil wars, ethnic and religious strife, and other forms of human conflict.

The key to effective poverty eradication and human development in the Nile River basin is for each riparian state to realize its full agricultural potential and for these countries to engage in the type of trade that would enhance specialization based on comparative advantage. In addition to the realization of the full potential of agricultural production in each riparian state, conditions that enhance cross-border trade, especially in agricultural products, must be created within the basin. Although many problems must be resolved before these countries can create the conditions that will enhance cross-border trade and significantly improve agricultural productivity, the development of irrigation systems is one of the most important. Such systems are needed to allow these countries to deal more effectively with the increasingly disruptive effects of climate change.

Many of the basin's governments have already developed irrigation master plans with emphasis on the provision of irrigation systems in particular and water infrastructure systems in general (NBI 2012, p. 155). Nevertheless, two factors are likely to represent important constraints to the effective development of irrigation systems in the Nile River basin: First, the development of irrigation systems requires enormous financial resources, which most of the Nile River riparian states do not have. Second, the water resources of the Nile River and its tributaries, which will form the source of the water needed for irrigated agriculture in the basin, are finite and must be shared by all eleven countries. Given that Egypt and the Republic of Sudan already claim more than 90 percent of the Nile River's waters, development of irrigated agriculture in the upstream riparian states would require creation of a legal framework for Nile River governance that replaces the *ancien* legal regime.[8] Hence there is a need to revisit the issue of a legal framework for Nile River governance.

8. The *ancien* or existing legal regime consists of the Nile Waters agreements—the 1929 Anglo-Egyptian treaty and the 1959 bilateral agreement between Egypt and Sudan. The upstream riparian states have indicated that they do not recognize these bilateral treaties and the rights created by them and that they are not bound by these agreements.

Climate-Change Water Stress in the Nile River Basin

Before we end this introductory section, it would be informative to briefly examine climate change and its implications for the basin generally and water use in particular. Climate change must be distinguished from climate variability. The United Nations defines climate change as "a change of climate which is attributed directly or indirectly to human activity that alters the composition of the global atmosphere and which is in addition to natural climate variability observed over comparable time periods." The Nile Basin Initiative "uses 'climate variability' . . . to refer to the intra-seasonal and inter-annual fluctuations in climatic parameters."[9] Although the Nile River remains the foundation for economic and human development in the eleven riparian states, the river has a relatively low supply of water, especially when compared with basins of comparable size (NBI 2012). Presently, there is a shortage of water to meet the basin's needs for commercial (for example, irrigation), domestic, and ecological projects. Thus water security is a serious problem, not only to Egypt but also to all the riparian states.

Today in the Nile River basin, food security is the most compelling policy issue. Hence food security, which is linked to and depends on adequate access to water, has emerged as the critical issue in Nile River negotiations.[10] As the basin's population increases, the demand for food, and hence water to produce it, is expected to increase significantly.[11] In recent years, it has become apparent that the riparian states will have to deal with another potentially devastating constraint to agricultural productivity: climate change. In fact, climate change has emerged as the driving force behind many of today's policy considerations regarding food and energy security.

During the next several decades, Africa is expected to be one of the regions most negatively affected by climate change. The region is susceptible to extreme weather events that are likely to produce famines, floods, and other weather-related calamities that could have significantly negative impacts on growth and development. Unfortunately, African countries presently do not

9. UN, United Nations Framework Convention on Climate Change, 1992 (http://unfccc.int/resource/docs/convkp/conveng.pdf); NBI (2012, p. 207). See also Downie, Brash, and Vaughan (2009).

10. More than 80 percent of the water withdrawn from the Nile River is used for irrigation, primarily to produce crops for both domestic consumption and export.

11. The populations of Egypt and Ethiopia are each expected to reach more than 100 million by the year 2025, and those of the other smaller riparian states are expected to increase significantly, easily approaching 60 million to 70 million per country.

have the capacity to deal effectively and fully with the effects of extreme weather events. For example, few Nile River basin countries have the necessary infrastructures and other mechanisms to deal with either floods or droughts. There are, of course, dams on the Nile and its tributaries that can be used to deal with these weather-related problems.[12] Nevertheless, "the capacity for soft tools like early warning systems for mitigation measures is . . . underdeveloped," and "there is a lack of long-period and good quality data and models to assess climate change and climate variability impacts in the region, so that so far, it has not been possible to undertake appropriate adaptive actions to mitigate climate change."[13]

The current challenge is for the Nile River basin countries to provide themselves with the necessary institutional mechanisms for mitigation and adaptation. Although the Nile Basin Initiative (NBI) has met climate-change issues head-on, it remains essentially a transitional body, as a permanent one is yet to be established. The Nile River basin countries have not been able to agree on a permanent legal framework for the governance of extraction and use of the waters of the river, further complicating the question of how to adequately deal with climate-change issues.[14]

The Nile River riparian states have prepared the National Adaptation Plans of Action, which identify priority areas for action by the basin as a whole and by individual countries. Flowing from these plans, are "specific policy targets and a list of practical climate-change adaptation projects for immediate implementation, for which funding is currently being sought" (NBI 2012, pp. 214–15). In addition, some of the riparian states have produced National Appropriate Mitigation Action programs. The government of Ethiopia, for example, has produced a Climate Resilient Green Economy initiative, which is building on the country's relatively low contribution to greenhouse gas emissions (NBI 2012, p. 214).

Mitigation programs can only function effectively if they are integrated into the country's development plans and civil society is fully engaged. Even taking current efforts into consideration, the "scope of the current climate

12. The Roseires Dam and Sennar Dam in the Republic of Sudan, Aswan High Dam in Egypt, and the Owens Fall Dam in Uganda.

13. "Impact of Climate Change on the Nile River Basin," 2014, University of Texas, Austin (www.ce.utexas.edu/prof/mckinney/ce397/Topics/Nile/Waako_etal_2009.pdf), p. 20.

14. The fight over the Nile Waters agreements and their relevance for governance in the Nile River basin remains a controversial issue and continues to consume the time of the governments of these countries. As a consequence, dealing with climate change has not been a priority with many riparian governments.

change adaptation programs and activities is not sufficient to deal with the scale of the threat, and these measures seem insufficient to prepare effectively for a changing climate."[15]

Climate change is expected to affect the Nile River basin in two principal areas: average temperatures and average precipitation in the basin. Temperature and precipitation, in turn, affect and to a certain extent control water levels in the region's wetlands, soils, rivers, lakes, springs, and groundwater aquifers (see, for example, NBI 2012, pp. 214–25). Specifically, the continuing warming trend is expected to have a significant impact on the Nile River basin: Many of the region's reservoirs are expected to lose significant amounts of water through evaporation. Drought risks will increase, especially during the dry season. Rainstorms are likely to be more frequent and severe in their intensity, significantly raising the risk of floods and damage to people and property. As a result of warmer temperatures, even higher altitudes are likely to become susceptible to vector-borne diseases such as malaria. And rising sea levels are likely to threaten both property and people in coastal communities (for example, Mombasa and Dar-es-Salaam). These impacts are likely to translate into higher levels of poverty and increased demand for environmental resources, including water (NBI 2012, p. 209).

Dealing effectively with climate-change issues must involve the development and implementation of sound economic and social policies, both at the regional and national levels. Adaptation and mitigation programs must be fully integrated into basin and national policy initiatives—stand-alone policies are not likely to function effectively to deal with climate change. Hence the Nile River basin riparians should opt for a regional approach to dealing with climate change. An appropriate place to begin would be the Nile Basin Initiative and its efforts to develop a new legal framework for the basin. However, since the initiative is a temporary institution, its work on climate change would eventually be handed over to the permanent regional institution that would be established to replace it.

Legal Issues Affecting the Nile River Basin

The Nile River is an international watercourse, shared by several countries. Like other shared watercourses, the Nile is governed by international water resources law. The latter may be derived from "international conventions, whether general or particular; . . . international custom, as evidence of a general

15. NBI (2014, chap. 8). See also UNEP (2012).

practice accepted as law; . . . the general principles of law recognized by civilized nations; and . . . judicial decisions and the teachings of the most highly qualified publicists of the various nations, as subsidiary means for the determination of rules of law."[16]

The most important method of international lawmaking has historically been the conclusion of international treaties. Thus the primary means of establishing rights and obligations over international watercourses has been the conclusion of treaties between the watercourse's riparian states. Such international treaties are necessary to provide for the reasonable utilization of the waters of the watercourse. However, for such treaties to create rights as between all watercourse riparians, all of these states must participate in the development of the treaty and become parties to it. Within the Nile River basin, however, such an international treaty does not currently exist. The closest thing to a legal regime for the management of the Nile River is a series of bilateral treaties agreed to among some riparian states as well as customary international law. Notable among these are the Nile Waters agreements—the 1929 Anglo-Egyptian treaty and the 1959 bilateral agreement between Egypt and Sudan.[17] Although there have been suggestions that these bilateral agreements or treaties reflect customary principles of law, that position has been challenged by both academics and policymakers in the upstream riparian states.[18]

The issue of determining rights and obligations over the waters of the Nile River rests on the legal status of the Nile Waters agreements. Put another way, is the "international legal regime established over the Nile [River]" through the treaties concluded between Great Britain and Egypt (1929) and between Egypt and Sudan (1959) still in force and binding on all the river's riparian states?[19] How this question is answered will determine the nature of the rights and obligations over the Nile River waters.

If the international legal regime established by the Nile Waters agreements is found to be valid and binding on all the Nile River riparian states, Egypt's

16. International Court of Justice, Statute of the International Court of Justice, art. 38 (www.icj-cij.org/documents/index.php?p1=4&p2=2&p3=0&#CHAPTER_II).

17. Exchange of Notes between His Majesty's Government in the United Kingdom and the Egyptian Government in Regard to the Use of the Waters of the River Nile for Irrigation Purposes (with Seven Diagrams), Cairo, May 7, 1929, L.N.T.S. 2103; United Arab Republic and Sudan Agreement (with Annexes) for the Full Utilization of the Nile Waters, Cairo, 8 November 1959, 6519 U.N.T.S. 63.

18. A. Okoth-Owiro, "The Nile Treaty: State Succession and International Treaty Commitments: A Case Study of the Nile Water Treaties," 2004, Konrad Adenauer Foundation, Nairobi, Kenya (www.kas.de/wf/doc/kas_6306-544-1-30.pdf).

19. Ibid., p. 13.

hegemonic hold on the basin would be legitimated and would severely constrain the ability of the upstream riparian states to engage in development projects (for example, irrigated farming, ecological restoration) that require access to the waters of the Nile River. However, if, as argued by the upstream riparian states, the Nile Waters agreements are not a valid and binding international legal regime for the management of the Nile River, then the development and utilization of the waters of the Nile are currently regulated only by customary international water law. This would imply that the Nile River basin can and should develop a basinwide legal regime, one that is compacted through an inclusive and democratic process. The development of such a new legal regime would provide riparians with the opportunity and wherewithal to bargain among themselves and reach an agreement that maximizes such critical values as sustainability and equity in the allocation and utilization of the basin's resources, as well as fairness.

The Riparian States and the Nile Water Agreements

Since the 1920s, Egypt has consistently asserted that the Nile Waters agreements are valid and binding on all the Nile River's riparian states. Egyptian authorities argue that these treaties can only be "amended or abrogated" by negotiation and in accordance with the provisions of the Vienna Convention on the Law of Treaties. In addition, Egypt has argued that treaties entered into by European powers on behalf of their colonies in Africa (for example, the 1929 Anglo-Egyptian treaty) remain valid and binding "by virtue of the law of state succession and because of the territorial nature of the obligations resulting from these treaties."[20] The Egyptians have also argued that they have acquired what they refer to as "natural and historic" rights (Mekonnen 2010, p. 432) to the Nile River through what essentially can be considered as the American "prior appropriation" doctrine (Getches 2009, p. 6): they claim that they have acquired these rights by their efforts to put the water to beneficial use over a long period of time and that these rights have been recognized by other countries, notably Great Britain and the Republic of Sudan. Postindependence Egyptian leaders have made access to the waters of the Nile River an issue of national security and have stated repeatedly that they would use force, if necessary, to prevent the diversion of water from the Nile River (see, for example, Collins 2002; Myers 1989; Starr 1991).

20. Ibid., "Nile Treaty," p. 16; see also Godana (1985).

Ethiopia's position regarding the Nile Waters agreements is based on the Harmon doctrine, which states that a state has absolute sovereignty over the water that flows through its territory and hence can utilize it in any way that it desires regardless of the impact on downstream riparians (McCaffrey 1996, 551–63; see also Kukk and Deese 1996–97). Ethiopian authorities state that they have absolute sovereignty over the portion of the Nile River that lies within their borders. They argue that they would not be breaking any laws if their use of the waters of the Blue Nile within their borders leaves no water to flow into the Nile River. In addition, Ethiopian leaders have argued that they are not bound by treaties entered into by Italy and other parties, supposedly on their behalf (Godana 1985).

Shortly after Tanganyika won independence from Great Britain, its government wrote letters to the governments of Great Britain, Egypt, and the Republic of Sudan stating that the country would not be bound by the Nile Waters agreements: "the provisions of the 1929 Agreement purporting to apply to the countries under British Administration are not binding on Tanganyika" (quoted in Howell and Allan 1994, p. 328). The letter also indicated that Tanganyika was willing to cooperate with other riparians with a view to developing and adopting a legal regime capable of enhancing the equitable and just allocation of the waters of the Nile River between all the river's riparian states (Seaton and Maliti 1974). The approach taken by Tanzania with regard to the Nile Waters agreements is embodied in what came to be known as the Nyerere doctrine, which states that "treaty relationships will be maintained for a limited period during which the new State will decide which treaties, both bilateral and multi-lateral, it wishes to continue indefinitely and which to discontinue altogether" (Schaffer 1981, p. 602; see also Collins 2002, p. 198).

In its postindependence period, Kenya adopted Tanzania's approach to treaty succession and, like Tanzania, argued that it was not bound by bilateral treaties concluded by Britain supposedly on behalf of the people of what was then the British colony of Kenya. Specifically, Kenya mentioned the 1929 Anglo-Egyptian treaty and the 1959 bilateral agreement between Egypt and Sudan. The country then declared its intention not to be bound by these treaties, and by communication to the secretary-general of the United Nations dated March 25, 1964, Kenya's prime minister indicated that the country would remain bound to colonial-era treaties for a period of two years from the date of independence (that is, until December 12, 1965) and that all treaties entered into on her behalf by the United Kingdom "may be abrogated or modified by mutual consent of the other contracting party before December 12, 1965."[21]

21. Quoted ibid., p. 39.

At independence, Uganda, Burundi, and Rwanda adopted a position on bilateral treaties, including the Nile Waters agreements, that mimicked the Nyerere doctrine (Okidi 1982; Brownlie 1990). After gaining independence in 1956, the Republic of Sudan declared its intention not to be bound by the 1929 Anglo-Egyptian treaty and forced Egypt into negotiations to create a new treaty, the 1959 bilateral agreement between Egypt and Sudan. The Democratic Republic of Congo has really never expressly declared its position on the Nile Waters agreements, while Eritrea has sided with Egypt. When it gained independence in 2011, South Sudan sided with the upstream riparian states and argued in favor of new negotiations to produce a more inclusive legal regime.

The foregoing discussion highlights the urgency of the need to revisit the Nile Waters agreements with a view to developing a new and more effective legal regime, specifically one that provides for equity and fairness in water allocation. The goal should be to explore the possibility of achieving Pareto-improving arrangements among the various states that make up the Nile River basin. Such welfare-enhancing arrangements can only be achieved through negotiations involving all the river's riparians.

3 *Setting the Stage for Conflict in the Nile River Basin*

Egypt, which boasts one of the oldest civilizations in the world, has sustained itself by extracting and using water from the Nile River (see, for example, Moret 2001). Over the centuries, Egyptians have come to claim "natural and historical rights" (Mekonnen 2010, p. 432) to its waters and resources.[1] Those claims have come to form the core of the conflict that modern Egypt has with upstream riparian states over the allocation and utilization of the waters of the Nile River. As other riparian states gained independence and developed the capacity to access more water from the Nile River and its tributaries, officials in Cairo came to consider these increased demands for water a threat to the security of the Egyptian state and its peoples. Of course, Egyptian governments have always considered the activities of Ethiopia, which was never a colony, on the Blue Nile as a threat to the survival of the country. Over the years, the authorities in Cairo have threatened to go to war to arrest

1. 1929 Anglo-Egyptian treaty (Exchange of Notes between His Majesty's Government in the United Kingdom and the Egyptian Government in Regard to the Use of the Waters of the River Nile for Irrigation Purposes [with Seven Diagrams], Cairo, May 7, 1929, L.N.T.S. 2103). This right, of course, is disputed by other Nile River riparian states, virtually all of which have condemned what they argue are attempts by Egyptian authorities to monopolize access to the waters of the Nile River and prevent the other riparians from accessing the waters for developing their own economies and meeting their national needs. The rights claimed by Egypt were granted by the Nile Waters agreements—the 1929 Anglo-Egyptian treaty and the 1959 bilateral agreement between Egypt and Sudan. Because the upstream riparian states have argued that they were not parties to these agreements and hence are not bound by them, the validity of the rights claimed by Egypt is in dispute.

any interference with the flow of water into the Nile River (see, for example, Klare 2002; Collins 2002; Mekonnen 2010, 2011). During the colonial period, several treaties were concluded between the states of the Nile basin to provide legal mechanisms for the allocation of the waters of the Nile River; most of these agreements were settled between Egypt and Great Britain, with the latter representing its colonies in the Nile River basin. These agreements, which continue to influence and significantly impact the governance of Nile River waters today, produced allocation regimes that favored Egypt and placed other riparians at a competitive disadvantage in regard to access to the waters and resources of the Nile River. Britain, which had superior power in the negotiations, favored Egypt in these agreements because it had significant agricultural interests in the Nile Delta and was determined to protect them.

Before the start of the Civil War in the United States, most of the cotton used by the world's textile mills was produced in the American South; a significant proportion of that cotton had been shipped to Lancashire in Great Britain to be woven into cloth (Beckert 2004, p. 1405). However, the outbreak of war in the United States in 1861 marked the beginning of what came to be known as the "cotton famine" (Earle 1926, p. 521). By 1862, the supply of American cotton to Lancashire had dropped precipitously to a small fraction of pre–Civil War amounts.

The cotton famine stimulated and provided the impetus for the transformation of Egyptian agriculture. Egypt lost its self-sufficiency in food production and became essentially a one-crop economy. That crop, of course, was cotton, which was critical to the British textile industry. In fact, by 1864 Egypt's agriculture was committed exclusively to cotton production. Given that the production of cotton in Egypt for export to Lancashire's mills was wholly dependent on water from the Nile River, the protection of the latter became a critical issue in British foreign policy, especially as regards the region. As already mentioned, in order to minimize interruption in cotton production, the British set out to develop, in Egypt, a policy of total dominance of the Nile (Earle 1926, pp. 521–22).

India had been expected to serve as an alternative source of supply of cotton to Lancashire, once the blockade of ports in the American South had gone into effect. However, India failed to supply cotton of comparable quality to American cotton. Egyptian cotton, however, equaled that of the American South, except for cotton that came specifically from Sea Island, Georgia. In addition, the output from Indian farms was not enough to meet the shortage created by the cotton famine (Earle 1926, p. 527). In replacing the United

States as the most important supplier of cotton to British industry, Egypt became a beneficiary of the American Civil War.[2]

Egyptian cotton provided a significant impetus to British imperial interests in both Egypt and the Nile River basin. When the American Civil War started, between 20 and 25 percent of all Britons were dependent, directly or indirectly, on the Lancashire mills, and nearly 80 percent of the cotton used in the latter came from the United States. By the early 1890s, the Nile River basin and its Egyptian cotton fields had become a significant contributor to the British economy. Thus Britain's decision to favor Egypt and Sudan in the various agreements regarding the management of the Nile River was based on a desire to safeguard its national interests.[3]

Britain consolidated its annexation of Egypt in 1882. From the start, the British had hoped that their administration of the captured territory would be short-lived and temporary. However, British authorities at Cairo believed that it was not enough to stabilize Egypt politically. They saw economic and financial stability as a critical part of the political stability of the territory and the region (Marlowe 1970; Baring 2010; Young and Young 2002). Hence at the behest of Evelyn Baring, first Earl of Cromer and the first controller general of Egypt, the British embarked on a program to strengthen Egyptian finances and modernize its economy. Part of the British restructuring of the Egyptian economy involved investing in what was then the most productive sector—cotton. Such investments necessarily brought the British directly into the business of regulating access to the waters of the Nile River.

Although British activities in Egypt, especially those associated with agricultural production, actually benefited Egyptians, many of the country's nationalists opposed colonization by Britain or any other European country. Thus in response to increased nationalist fervor for independence, the British unilaterally declared Egypt independent, effective February 28, 1922.

Today, the conflict in the Nile River basin involves four issues: how to efficiently and sustainably manage the basin's resources, including its waters; how to allocate the basin's resources, especially its waters, among all the riparian states in a reasonable and equitable manner; how to deal with the Nile Waters agreements, which Egypt and the Republic of Sudan consider to be the basin's binding legal framework but which the upstream riparian states,

2. Cotton production in the American South declined after the Civil War ended. With the slaves freed, cotton growers were now forced to pay wages to those they employed to work in the cotton fields, a development that significantly increased the cost of production of cotton in the American South. See, generally, Cohn (1956).

3. Plusquellec (1990).

for various reasons, do not recognize as binding on them; and how to produce a new legal framework that is acceptable to all of the Nile River basin's riparian states and communities. Conflict over these issues continues to impede political and economic cooperation in the Nile River basin.

Upstream riparian states have denounced the Nile Waters agreements and want them either renegotiated or totally abrogated and a new, more comprehensive, inclusive, and relevant legal and institutional arrangement established.[4] Specifically, the upstream riparian states want a legal regime in the Nile River basin that enhances and guarantees equitable and fair allocation of the water and resources of the Nile River and its tributaries.[5] This conflict can be more fully and effectively evaluated and understood in the context of the region's colonial history (Kendie 1999).

Between November 15, 1884, and February 26, 1885, several European countries met in Berlin to find ways to bring some order to the way they were exploiting Africa's enormous resources. The Berlin conference produced an agreement on how the European colonizers would divide Africa among themselves. By the time World War I started in 1914, most of Africa had come under the effective control of several European countries, notably, Great Britain, France, Portugal, Spain, Belgium, and Germany. These countries restructured the territories that they had captured into colonies, which were expected to serve as critical markets for excess output from metropolitan factories and sources of raw materials for European industrial production (see, for example, Lugard 1926; Egerton 1969; Webster 2006).

In the language of the colonial period, the African territories were the colonies, and each European country was the "metropole" (Webster 2006, pp. 68–92). The colonies were established through force, and each usually consisted of an aggregation of several ethnic and religious groups. There was

4. The Nile Waters agreements are the 1929 Anglo-Egyptian treaty and the 1959 bilateral agreement between Egypt and Sudan (United Arab Republic and Sudan Agreement [with Annexes] for the Full Utilization of the Nile Waters, Cairo, 8 November 1959, 6519 U.N.T.S. 63).

5. For example, in 2002, Kenya's energy minister, Raila Odinga, denounced the Nile Waters agreements, calling them "obsolete" and irrelevant to Kenya's development agenda (John Kamau, "Can EA Win the Nile War?," *Nairobi Daily Nation,* March 28, 2002). The agreements were deemed irrelevant because they were negotiated without the participation of all relevant stakeholder groups—especially the upstream Nile states. Hence, they fail to reflect these stakeholders' view of how the Nile River's resources should be developed and utilized. A legal regime, in this view, would take into account the specificities of not just each of the eleven countries but also the various communities that inhabit the river's banks.

virtually no consultation with each colony's relevant stakeholders. Although in some rare cases there may have been some agreements that were based on mutual benefit, the African political or traditional elites who signed these treaties of annexation were usually not effective representatives of the interests of their populations. Although such agreements produced significant benefits for the African political and traditional leaders who cooperated with the colonizers (and, of course, their European counterparts), they imposed significant costs on the majority of Africans—including loss of access to their environmental resources generally and their land specifically (Rudin 1938; Burns 1963).

That the Europeans preferred force and coercion in their relations with Africans, especially in the acquisition of land, is undeniable. In fact, many colonial officers freely expressed their belief that the most effective way for them to resolve conflicts with various African groups and maximize their interests was to rely on instruments of coercion. Consider the proclamations of a distinguished British colonial officer in southern Africa, Earl Grey, when he stated that whenever and wherever there was a conflict between the European colonists (settlers) and indigenous African groups, the interests of the latter should be ignored and "facilities should be afforded the white colonist for obtaining the possession of land theretofore occupied by the Native tribes" (quoted in Magubane 1979, p. 71).

British colonial officers were not the only ones who relied on force as the principal method through which they engaged with Africans. In fact, all the European countries that were involved in the "scramble for Africa" (Pakenham 1991) in the mid to late 1800s employed virtually the same tactics, all of which were characterized by a failure to fully and effectively engage Africans in efforts either to set up the colonies or to manage them. As stated by then French governor of Algeria, General Bugeaud, "Whenever the water supply is good and the land fertile, there we must place colonists without worrying about previous owners. We must distribute the lands [with] full title to the colonists" (Brace 1964, p. 48).

To enhance and strengthen their ability to exploit Africa's resources for their own benefit and that of their economies back in Europe, colonial officers established two important regimes in each colony that were to significantly impact the allocation of water in the postindependence period. First, using their superior military and police power, they brought together various ethnic groups, each with a distinct language, culture, and set of customs as well as social, political, and economic systems, to form a political and economic unit that could only be held together through the use of a significant level of coercion (Morrison, Mitchell, and Stevenson 1989). Second, the colonialists imposed their laws and institutions on the colonies. These institutional arrangements were not designed to enhance the ability of Africans to efficiently govern themselves and allocate their resources in a productively and allocatively efficient manner. Instead, the legal and judicial systems that were established in the colonies were intended to provide the colonial government the wherewithal to subdue Africans and allow colonists to engage in various forms of economic exploitation (Burns 1963; Rudin 1938).

According to Robert Fatton Jr. (1990, p. 457), the laws and institutions that the various European countries imposed on the African colonies were primarily "structures of exploitation, despotism and degradation" (see also Crowder 1987, pp. 11–12). Michael Crowder argues that during the colonial period, the Europeans ruled with brute force. The colonial state, Crowder states, "was conceived in violence rather than by negotiation. This violence was often quite out of proportion to the task in hand, with burnings of villages, destruction of crops, killing of women and children, and the execution of leaders. . . . The colonial state was not only conceived in violence, but it was maintained by the free use of it. Any form of resistance was visited by punitive expeditions that were often quite unrestrained by any of the norms of warfare in Europe" (Crowder 1987, pp. 11–12). Since the colonial state was created through violence and not through mutually beneficial arrangements, it was inevitable that the laws and institutions that served as mechanisms for administering the colonies and maintaining the coexistence of each colony's diverse population groups were themselves instruments of violence.

These colonial institutional arrangements included treaties and international conventions and covenants concluded by the colonial powers on behalf of the colonies and supposedly designed to protect the interests of the latter's inhabitants. Examples include the 1891 Anglo-Italian protocol, in which Britain represented colonial Egypt and Sudan and Italy acted on behalf of Eritrea;[6] agreements between a colonial power (which usually enjoyed a comparative advantage in the employment of military force and hence boasted a more favorable negotiating position) and an independent African country,

6. Protocols between the Governments of Her Britannic Majesty and the King of Italy, for the Demarcation of Their Respective Spheres of Influence in Eastern Africa, Signed at Rome, March 24 and April 15, 1891, 1890–91 [c. 6316] XCVI.383; Treaties between the United Kingdom, Italy, and Ethiopia, Relative to the Frontiers between the Soudan, Ethiopia, and Eritrea, Signed at Addis Ababa, May 15, 1902, Treaty Series No. 16 (1902); and Agreement between the United Kingdom, France, and Italy, Respecting Abyssinia, signed at London, December 13, 1906.

such as the 1902 accord between Britain, representing the colony of Sudan, and an independent and sovereign Ethiopia; and tripartite treaties between European countries governing the allocation of some important African environmental resource, such as the 1906 treaty involving Great Britain, France, and Italy on the allocation of the waters of the Nile River, with specific reference to the Ethiopian subbasin.

These agreements shared certain important characteristics: Relevant stakeholders, that is, those African population groups whose livelihood and welfare were to be directly affected by the allocation of the resource in question, were not provided the facilities to participate fully and effectively in the design of the agreements. The main objective of the agreements was to safeguard the interests of the European power of record, and little consideration was given to the needs, desires, traditions, customs, and cultures of the indigenous groups in whose lands the resource was located; in fact, issues such as resource sustainability and ecosystem degradation, which were important to the local constituencies, were usually ignored when these agreements were designed.[7] Even in the case where a colonial government was negotiating or acting on behalf of an African population, the interests of that population were rarely taken into consideration or given an appropriate hearing. This is evident, for example, in British attitudes toward the allocation of the waters of the Nile River. Here, Britain represented Egypt and various colonies, which were Nile River riparians (for example, Sudan, Tanganyika, Uganda, Kenya), but favored an allocation formula that reserved virtually all of the waters of the Nile River to Egypt.

Britain's desire to safeguard its agricultural and industrial interests in Egypt is evident in the 1902 Anglo-Ethiopian treaty. In 1902, the British government sent Lieutenant Colonel John Harrington to Addis Ababa to consult with Ethiopia's emperor Menelik II regarding British interests in the waters of the Nile River. According to article 3 of the treaty, which was signed on May 15, 1902, "His Majesty the Emperor Menelik II, King of Kings of Ethiopia, engages Himself towards the government of His Britannic Majesty not to construct or allow to be constructed any work across the Blue Nile, Lake Tsana or the Sobat, which would arrest the flow of their waters into the Nile except in agreement with His Britannic Majesty's Government and the Government

7. To be fair, it should be noted that, owing to ignorance and the absence of supporting research, few nations at the time seriously considered issues of resource sustainability and ecosystem degradation to be critical. Nevertheless, that these issues were not adequately addressed remains relevant and important and should be part of the historical examination of the basin's political economy.

of the Soudan." Britain signed that agreement on behalf of Egypt and Sudan, which at the time were British colonies, administered and governed by London. Britain, which had large cotton plantations in Egypt, made Egypt's interests its own.

Many colonial accords on the allocation of the waters of the Nile River sought to prevent upstream riparian states from engaging in activities (for example, construction of diversion schemes) that might jeopardize Egypt's ability to secure all the water it needed for its agricultural and industrial activities. The 1925 Anglo-Italian agreement specifically stated that the British government's recognition of Italian interests in the region was "subject to the proviso that the Italian Government on their side, recognizing the prior hydraulic rights of Egypt and the Sudan, will engage not to construct on the head waters of the Blue or White Niles or their tributaries or affluents any work which might sensibly modify their flow into the main river."[8]

Several upstream riparian states have warned that unless the Nile Waters agreements are renegotiated through a participatory and inclusive process or a multilateral legal framework is created, violent conflicts are likely to engulf the region. The relationship between Egypt and Ethiopia is especially critical to the maintenance of peaceful coexistence in the Nile River basin. Unfortunately, the probability of violent confrontation between the two countries is quite high, especially given that Egypt continues to insist that Ethiopia is bound by the Nile Waters agreements, neither of which Ethiopia is a party to (Kendie 1999, pp. 145–47).

Many of the Nile River basin's upstream riparian states believe that the present water allocation regime is unfair, inequitable, and unsustainable.[9] The Nile Waters agreements allocate most of the waters of the Nile River to Egypt and Sudan. In addition, they significantly constrain the upstream riparian states in their ability to use the Nile River for local economic development. During the past several decades, there has been a significant increase in the demand for water among the Nile River basin states, owing not only to rapid population growth but also to recent increases in economic activities, significant improvements in the ability of many of these states to utilize

8. Exchange of Notes between the United Kingdom and Italy Respecting Concessions for a Barrage at Lake Tsana and a Railway Across Abyssinia from Eritrea to Italian Somaliland, December 14 and 20, 1925, No. 1 & No. 2, L.N.T.S. 1211. The quote comes from the letter dated December 14, 1925.

9. Politicians in Kenya have openly called for the country to ignore the agreements and proceed with its plans to tap the waters of Lake Victoria for development. See, for example, Kamau, "Can EA Win the Nile War?"

the waters of the Nile River for productive activities, and the impacts of climate change. Unfortunately, no institutional framework currently exists within the basin that can provide the legal structure for the efficient, equitable, fair, and sustainable management of the Nile River's waters (see, for example, Knobelsdorf 2005–06).

4

The Nile Waters Agreements: A Critical Analysis

In many societies, the allocation of water and the other resources (for example, the rights to remove and utilize fish or use the river for navigation) of a watercourse has historically been governed by custom and tradition. However, as these societies have become more complex, they have relied on more explicit rules for regulation of the use of water resources. These rules include national laws and institutions and international conventions, protocols, and covenants. However, where access to water resources is governed by written rules, tradition and custom still remain important and continue to influence access to and use of these scarce resources.

The governing of the allocation and use of the waters of the Nile River, like that of other watercourses around the world, is currently undertaken through custom and tradition, local written laws, and international conventions. In this chapter, we provide a more in-depth analysis of the various international conventions that have been designed specifically to deal with conflict over the allocation and use of the waters of the Nile River. Such an analysis is critical to understanding the need for the Nile River basin countries to develop a new and inclusive legal framework for governing the basin and its water resources.

Before the twentieth century, the Nile River downstream riparians—Egypt and Sudan—were more interested in "controlling floods along the Nile than with staking a legal claim to an increasingly scarce resource" (Knobelsdorf 2005–06, p. 626). The upstream riparians, all of which, except Ethiopia, were colonies, did not engage in any development projects (such as the massive irrigation agriculture that occurred in Egypt) that put the waters of the Nile

River to systematic use. This was primarily because these colonized communities did not have the technological and economic capacity to engage in large irrigated agricultural projects. However, since the end of colonialism, many of the upstream riparian states, including Kenya, Tanzania, and Uganda, have embarked on development programs that require more consistent and reliable access to the waters of the Nile River. Many of these development plans have become viable as these countries have acquired the necessary capacity to implement such projects. In addition, increased population pressures and climate change have made conflict among the various riparian states over access to the waters of the Nile River inevitable (Knobelsdorf 2005–06, p. 626; Carroll 1999, pp. 270–71).[1]

Other problems have contributed to the ongoing conflict in the Nile River basin. These include chronic droughts and floods, which have made managing the Nile River's waters and its resources extremely challenging; climate change, which is said to be contributing significantly to reduction in the flow of water into the Nile River; and significant increases in the demand for food, whose production can no longer be carried out fully through rain-fed farming processes. Recently, Stanislas Kamanzi, the minister for environment and land in Rwanda, an upstream riparian, stated that because of climate change, Rwanda can no longer depend exclusively on rain for its agricultural production but must now seek to make more effective use of the waters of the Nile River basin to support a more modern and scientific approach to food production. According to the minister, "Everyone knows we are all faced with the realities of climate change." He continued that "Rwanda is one of the most vulnerable countries, 80 percent of our agriculture relies on rain-fed irrigation, and we can no longer predict that we will receive regular rainfall, so we can't predict our crop production. We need to use the resources in our waterways and lakes, and these are clearly in the Nile Basin."[2] Rwanda, Uganda, and Ethiopia are upstream riparian states that have, in the past, experienced significant and brutal ethnic conflict. These countries now have more stable political and economic systems and are currently able to attract significant investment from both internal and external sources. Part of that investment is directed at exploiting opportunities in commercial agriculture. The success of the latter depends not on rain-fed agriculture but on effective exploitation

1. By 2025, it is estimated, the population of the Nile River basin will reach 400 million people. See, for example, Carroll (1999, p. 275).

2. Quoted in Mike Pflanz, "Egypt, Sudan Lock Horns with Lower Africa over Control of Nile River," *Christian Science Monitor*, June 4, 2010, p. 2.

of the waters of the rivers and lakes that supply the Nile River. These developments raise concerns among Egyptians and Sudanese, who continue to insist that, based on the Nile Waters agreements, no upstream riparian state has the legal right to access Nile River–related waters without their express permission.

These and other problems have made efficient and equitable management of the waters of the Nile River crucial and necessary to "ensure food security, public health, economic growth, and maintenance of natural systems" (Carroll 1999, p. 275). The Nile River basin countries must seek ways to peacefully resolve water-related conflicts in order to create the necessary environment for cooperation in dealing with the multifarious problems that they face.

Today, all the upstream Nile River riparian states openly criticize the Nile Waters agreements, which, they argue, have effectively deprived them of reasonable and equitable access to the Nile River's waters.[3] Egypt, on the other hand, steadfastly and stubbornly adheres to what is essentially a "nationalist theory" of territorial water rights, "according to which all important works on the Nile should be constructed in Egyptian territory in order to avoid the danger of any works built outside of the country being used as a political weapon against Egypt" (Caponera 1993, p. 653). Over the years, Egypt has, on many occasions, demanded that Ethiopia not build any structures on the Blue Nile or on Lake Tana, its source, that might interfere with the flow of water into the Nile River (Kendie 1999, p. 141).

Most of the water that flows into the Nile River and is available for use in Egypt comes from the Blue Nile and the White Nile, which originates from Lake Victoria in Uganda (Kendie 1999, p. 141). Throughout history, it has been argued that "Egypt must be in a position either to dominate Ethiopia, or to neutralize whatever unfriendly regime might emerge there" (Kendie 1999, p. 141), as well as do whatever was necessary, including going to war, to ensure that no manmade action interfered with the flow of water into the Nile River. In fact, the late Egyptian president Anwar el-Sadat has argued that "any action that would endanger the waters of the Blue Nile will be faced with a firm reaction on the part of Egypt, even if that action should lead to war" (Wright 1972, p. 44).

According to an observer of Egyptian political economy,

> Egypt is a country that has not abandoned its expansionist ambitions. It regards its southern neighbors as its sphere of influence. Its strategy

3. See generally John Kamau, "Can EA Win the Nile War?," *Nairobi Daily Nation*, March 28, 2002.

> is essentially negative: to prevent the emergence of any force that could challenge its hegemony, and to thwart any economic development along the banks of the Nile that could either divert the flow of the water, or decrease its volume. The arithmetic of the waters of the Blue Nile River is, therefore, a zero-sum game, which Egypt is determined to win. It must have a hegemonic relationship with the countries of the Nile Valley and the Horn of Africa. When, for instance, Ethiopia is weak and internally divided, Egypt can rest. But when Ethiopia is prosperous and self-confident, playing a leading role in the region, Egypt is worried.[4]

In response to the above statements, the Egyptian ambassador to Ethiopia, Marawan Badr, stated as follows: "Such political commentary, or more correctly, political trash, cannot come [except] from a sick and disturbed mind. Egyptian-Ethiopian relations are not in a crisis. We do not even have problems. There are serious issues, which need to be addressed" (quoted in Kendie 1999, p. 142).[5]

It is possible that Badr was trying to be diplomatic when he claimed that there was no crisis in Egypt's relations with Ethiopia. Since the Blue Nile, which lies within the Ethiopian highlands, supplies as much as 85 percent of the water that flows into the Nile River, it is actually the backbone of Egyptian agriculture, industry, and livelihood. Yet over the years Egyptian authorities have continued to threaten Ethiopia against making use of the Blue Nile for Ethiopia's own national development (Wright 1972, p. 44).

One contributor to the potential conflict that currently afflicts the Nile River basin is the failure of the riparian states to design and adopt an effective legal framework, one that is acceptable to all of them. Associated with that conflict is that the upstream and downstream riparians have not been able to agree on what role, if any, the Nile Waters agreements should play in Nile River governance. These agreements continue to color any discussion of Nile River governance and render cooperation between all riparian States relatively difficult.

4. Quoted in Kendie (1999), p. 141. The statement was originally made in the *Addis Tribune* (Addis Ababa, Ethiopia), June 26, 1998, in an article titled "Ethiopia and the Horn of Africa."

5. The ambassador's statement originally appeared as "Egypt and the Horn of Africa: The True Perspective," *Addis Tribune* (Addis Ababa, Ethiopia), August 14, 1998. Although this could be considered a biased opinion piece provided by Ethiopians who want to demonize Egypt and make the country the "bad guys" in this conflict, it is still useful because it provides insight into how some Ethiopians feel about the continuing struggle of their country to deal with Nile River–related issues.

The Nile Waters Agreements

The Nile Waters agreements consist of two treaties. One is the Anglo-Egyptian treaty of 1929, which allocated the waters of the Nile River between Egypt and Sudan. The other is the 1959 bilateral agreement between Egypt and Sudan, which made adjustments to the allocations established by the 1929 agreement in response to complaints from an independent Sudan that it needed more water for its economic development. Below, we undertake a more in-depth analysis of the two treaties.

The 1929 Anglo-Egyptian Treaty

The 1929 Anglo-Egyptian treaty is an exchange of notes between the government of the United Kingdom, representing its various colonies in the Nile River basin, including Sudan, and Egypt.[6] The treaty consists of a letter from the government of Egypt to the government in the United Kingdom and the 1925 report of the Nile Commission, with its several appendixes. In the letter, the parties recognize that "the development of the Sudan requires a quantity of the Nile water greater than that which has been so far utilized by the Sudan" (Anglo-Egyptian treaty, para. 2). The Egyptians, however, caution that though the Egyptian government was willing to agree to efforts to grant the colony of Sudan additional water from the Nile River, it would undertake that task only if doing so "does not infringe Egypt's natural and historical rights in the waters of the Nile" (para. 2). Later portions of the letter are devoted to issues of "irrigation works" on the Nile River (para. 4). A subparagraph places constraints on upstream riparians regarding their ability to construct structures on the Nile River or its tributaries:

> Save with the previous agreement of the Egyptian Government, no irrigation or power works or measures are to be constructed or taken on the River Nile and its branches, or on the lakes from which it flows, so far as all these are in the Sudan or in countries under British administration, which would, in such a manner as to entail any prejudice to the interests of Egypt, either reduce the quantity of water arriving in Egypt, or modify the date of its arrival, or lower its level (Anglo-Egyptian treaty, para. 4[ii]).

6. The 1929 Anglo-Egyptian treaty is officially referred to as Exchange of Notes between His Majesty's Government in the United Kingdom and the Egyptian Government in Regard to the Use of the Waters of the Nile River for Irrigation Purposes (with Seven Diagrams), Cairo, May 7, 1929, L.N.T.S. 2103 (1929).

The 1925 Nile Commission Report, which is part of the treaty, consists of five chapters and several appendixes. Chapter 1 discusses the Nile Commission, how it was constituted, how its members were appointed, and a brief mention of its work. Chapter 2 is devoted to an examination of the history of previous irrigation projects on the Nile River and its tributaries. Following the overview, attention is then directed at the elaboration of projects proposed for the Nile River—referred to as the "present position" (Anglo-Egyptian treaty, chap. 2). Of special note is article 22, which states that "the British Government, however, solicitous for the prosperity of the Sudan, have no intention of trespassing upon the natural and historic rights of Egypt in the waters of the Nile, which they recognize to-day no less than in the past."

Chapter 3 of the treaty examines the relevant Nile River–related statistics; there is a discussion of "the nature of the records available" and "certain factors affecting the calculations" (chap. 3, art. 42). The treaty states that the hydrological records are available "from 641 to 1451 A.D. and again from 1737, with one break, to the present day" (chap. 3, art. 48). Detailed information is provided not only on the Nile River but also on its tributaries, including the Blue and White Niles.

Chapter 4 is devoted to an examination of basin and pump irrigation in the Sudan. It is noted that at the time of the treaty, irrigation was not a significant part of agriculture in the Sudan, since only a small amount of acreage was then devoted to irrigated agriculture. This section also discusses the possibility of increasing the area under pump irrigation in the Sudan.

Chapter 5 provides a summary of the commission's findings. The highlights of the summary and conclusions include the following: a statement that the natural flow of the Nile River "should be reserved for the benefit of Egypt from the 9th January to the 15th July (at Sennar), subject to the pumping in the Sudan"; a statement that the Gezira Canal, in Sudan, under conditions elaborated in the treaty, was allowed to draw on the natural flow of the river beginning on July 16, 1929; an enumeration of specific volumes of water to be drawn by the Gezira Canal from August 1 to December 3, 1929; and other factors affecting the withdrawal of water from the Nile River and its tributaries by the Gezira Canal.

When the negotiations for the agreement were being conducted, Ethiopia was not a British colony. Ethiopia, or the Abyssinian Empire, was an independent sovereign polity at the time the 1929 agreement was concluded. Hence, Ethiopia was neither a signatory, directly or indirectly, to the 1929 Anglo-Egyptian treaty nor a participant in the negotiations that led to its conclusion. Ethiopian government officials have since refused to recognize the legality of the 1929 agreement, arguing that the country had never been colonized by Britain and had never participated in putting together the agreement. As stated by Kendie (1999, p. 147), Ethiopian authorities considered the agreement to be grossly unfair and inequitable since "one party [Egypt] reserved for itself all the rights and privileges, leaving the other party [the upstream riparian states] without any quid pro quo." Again, according to Kendie (1999, p. 147), Ethiopian officials also argued that "the whole exercise of the agreement was geared mainly to protect and to promote Egypt's interests without any reciprocity, and that it had not renounced its own quantitatively unspecified but existing natural right to the Nile waters in its territory." In fact, as far back as the mid-1950s, Ethiopia had asserted and reserved its rights to access waters of the Blue Nile for national development without worrying about restrictions supposedly imposed on Ethiopia by Egypt.

1959 Bilateral Agreement between Egypt and Sudan

The 1959 bilateral agreement between Egypt and Sudan effectively reinforced the provisions of the 1929 Anglo-Egyptian treaty and provided for "full utilization of the Nile Waters."[7] Letting the Sudd el Aali Reservoir be the point for measuring the average water yield, the 1959 treaty increased allocations to both Egypt and Sudan—Egypt's allocation was increased from 48 billion cubic meters to 55.5 billion cubic meters and Sudan's from 4 billion cubic meters to 18.5 billion cubic meters.[8] These allocations were based on an estimated average annual yield at Aswan of 84 billion cubic meters. The treaty left 10 billion cubic meters to account for seepage and evaporation. In the case of an increase in average yield, the treaty dictates that the increase be shared equally between the two countries.

The 1959 agreement begins by laying out the main objective of the treaty: to make allowance for the full utilization of the waters of the Nile River by Egypt and Sudan. For example, the introduction states that "as the Nile waters agreement concluded in 1929 provided only for the partial use of the Nile waters and did not extend to include a complete control of the River waters, the two Republics have agreed on the following" and continues to delineate what are referred to as "the Present Acquired Rights." These rights are defined as 48 milliards (billion) cubic meters a year of waters from the Nile River for Egypt and 4 milliards cubic meters a year for Sudan at Aswan (art. 1, para. 1).

7. United Arab Republic and Sudan Agreement (with Annexes) for the Full Utilization of the Nile Waters, Cairo, November 8, 1959, 6519 U.N.T.S. 63.

8. Agreement of 1959 between Egypt and Sudan, arts. 1 and 2.

Article 2 deals with Nile control projects and the division of their benefits between Egypt and Sudan. This section of the treaty grants Egypt permission to use the Sudd el Aali reservoir at Aswan to enhance water storage and regulate and control the Nile River's waters. This is just one of several projects that the treaty anticipated would be constructed on the Nile River and its tributaries to enhance water management and use (art. 2, para. 1). To improve Sudan's ability to harvest and utilize the waters of the Nile River, the treaty grants Sudan permission to construct the Roseires Dam on the Blue Nile and "any other works which the Republic of the Sudan considers essential for the utilization of its share" (art. 2, para. 2).

The "net benefit from the Sudd el Aali Reservoir," the treaty states, "shall be calculated on the basis of the average natural River yield of water at Aswan in the years of this century, which is estimated at about 84 Milliards of cubic meters per year" (art. 2, para. 3). This flow of water would be shared by the two countries as indicated in the treaty. The first article of the treaty recites the allocation made possible by the 1929 Anglo-Egyptian treaty and elaborates on those allocations.

This section of the treaty makes allowance for Egypt to compensate Sudan "15 Million Egyptian Pounds" for damage done to Sudanese properties by storage at the Sudd el Aali Reservoir (art. 2, para. 6). In addition, the government of Sudan is instructed to evacuate all residents whose lands and properties would be "submerged by the stored water" (art. 2, para. 7).

Article 3 is devoted to an explication of "projects for the utilization of lost waters in the Nile Basin." Article 4 deals with "technical cooperation between the two republics." The two parties—Egypt and the Republic of Sudan—agreed that after the signing of the agreement, a permanent Joint Technical Commission, consisting of "equal members from both parties," would be formed to perform specific functions elaborated in the treaty (art. 4, para. 1). In addition to supervising "the execution of the projects approved by the two Governments" (art. 4, para. 1[b]), the Joint Technical Committee was also "charged with the task of devising a fair arrangement for the two Republics to follow" (art. 4, para. 1[e]) in the case of low levels of water in the Sudd el Aali Reservoir.

The fifth article is devoted to "general provisions," many of which have a significant impact on the treatment of the upstream riparian states. For example, article 5, paragraph 1 states as follows:

> If it becomes necessary to hold any negotiations concerning the Nile waters, with any riparian state, outside of the boundaries of the two

> Republics [Egypt and the Republic of Sudan], the Governments of the Sudan Republic and the United Arab Republic shall agree on a unified view after the subject is studied by the said Technical Commission. The said unified view shall be the basis of any negotiations by the Commission with the said states.

To strengthen their ability to control the Nile River, the two signatory states to the 1959 compact agreed to act with one voice in facing any challenges posed by other riparian states to the allocation of the waters of the Nile River (art. 5, para. 1). As mentioned earlier, the Joint Technical Commission was expected to make sure that if any upstream riparian state wanted to construct any structures on the Nile River, it had to obtain the permission of the two states parties to the 1959 agreement and subject the project to mandatory oversight and supervision by the commission. Specifically,

> If it becomes necessary to hold any negotiations concerning the Nile waters, with any riparian state, outside the boundaries of the two Republics, the Governments of the Sudan Republic and the United Arab Republic shall agree on a unified view after the subject is studied by the said Technical Commission. The said unified view shall be the basis of any negotiations by the Commission with the said states.
>
> If the negotiations result in an agreement to construct any works on the river, outside the boundaries of the two Republics, the joint Technical Commission shall after consulting the authorities in the Governments of the States concerned, draw all the technical execution details and the working and maintenance arrangements. And the Commission shall, after the sanction of the same by the Governments concerned, supervise the carrying out of the said technical agreements. (Art. 5, para. 1)

The Joint Technical Commission was granted the power to supervise all projects on the Nile River, including those outside the two republics. The commission was also required by the treaty to consult regularly with the two riparian states to make sure that their water use did not exceed the allocations granted them by the treaty.

Article 6 deals with the transition period before the Sudd el Aali is completed and becomes functional. Egypt and Sudan were required by the treaty to agree on a joint policy on water utilization for the transition period—that is, the time between the conclusion of the treaty and the completion of the Sudd el Aali.

The rest of the treaty, including annexes 1 and 2 and a protocol, deals with various issues between the two republics and the other riparians, including a section elaborating provisions of the treaty. The protocol to the treaty provides details on the establishment of the permanent Joint Technical Committee. Article 3 of the protocol provides that "the present protocol shall be considered supplementing the Agreement for the Complete Utilization of the Nile Waters signed on November 8, 1959, and shall have effect as of the date of its signing."

The question is whether the Nile Waters agreements restrict the rights of the upstream riparian states to the use of the waters of the Nile River. The upstream riparian states denounced these treaties and argued at various times that they are not bound by them. If there is any binding effect, it would most likely come from the imbalance in military and economic power between Egypt and the upstream riparians and not from the treaties themselves. If the upstream riparian states ignore the treaties, as Ethiopia has done in its decision to move forward with the construction of the Grand Ethiopian Renaissance Dam, Egypt's recourse would have to be either legal or military. The former is most likely to be considered appropriate by the international community; the latter would be viewed as inappropriate. Short of military action, how would Egypt and Sudan enforce this bilateral agreement against nonsignatory states? What if Ethiopia, for example, ignored the 1959 agreement and built or attempted to build structures on the Blue Nile or on Lake Tana that effectively decreased water flows from the Blue Nile to the Nile River?

Ethiopian-Egyptian Relations

Egyptian foreign policy has, for many years, been shaped by "the hydropolitics of the Nile in general and the Blue Nile in particular" (Kendie 1999, p. 151). From 1865 to 1885 Egypt controlled the port of Massawa, as well as "parts of present-day northwestern Eritrea from 1872 to 1884, with a view to using these areas as bases for military operations against the rest of Ethiopia" (Kendie 1999, p. 151; see also Degefu 2003; Milkias and Metaferia 2005; Jonas 2011). Egypt used its occupation of lands that were to become parts of present-day Eritrea to justify its argument that it had a legitimate historical right and legitimacy to enforce its hegemony on the region and that as "early as 1945 Egypt instigated the Arab League to declare its intention to put Eritrea under the Trusteeship of the Arab nations" (Kendie 1999, p. 151).

When Gamal Abdel Nasser came to power in Egypt in 1956, he initiated a campaign to bring together various Nile River basin communities under

Egyptian control. However, several developments in the region frustrated his plans. These included the reunification of Eritrea with Ethiopia in 1952, the independence of Anglo-Egyptian Sudan in 1956, and the emergence of an independent Somalia in 1960 (Kendie 1999, p. 153).[9]

While serving as an army officer in Sudan, Nasser had made contacts with the Ethiopian ruler Emperor Haile Selassie. After he became the leader of Egypt in 1956, Nasser extended several invitations to the Ethiopian leader to visit Egypt, but the King of Kings of Ethiopia declined.[10] Although Nasser continued to shower both Ethiopia and its leader with compliments and maintained his campaign to persuade the Ethiopian leader to visit Egypt, "Ethiopian functionaries were filled with fear as they watched Nasser undermine the monarchical regimes in Jordan, Iraq, Saudi Arabia, and Libya" (Erlich 1994, p. 133). In December 1956, Haile Selassie sent Meles Andom, the Ethiopian ambassador to Sudan, to Cairo to meet with Nasser and explain why the emperor had declined his various invitations to visit Egypt. After his return from Cairo, Andom described his meeting with the Egyptian leader to the British ambassador:

> Nasser had then asked whether a military alliance between Egypt, the Sudan and Ethiopia would not be in their best common interest. "We drink of the same water," he said. My Ethiopian colleague had replied bluntly, to the following effect: You claim to be an Arab and to lead the Arab world but you interfere in the affairs of your Arab neighbors and have tried to cause trouble for the Governments of Iraq, Libya, Lebanon and Sudan. We Ethiopians are not Arabs. We are Africans and we are black. We do not belong to your world although like you we drink of the water of the Nile. Yet you have tried to interfere in our affairs also and make trouble for His Majesty. . . . Secondly, you have military objectives. We do not know exactly what they may be but we have no confidence in the strength of your armed forces, and we are strongly against the Communists who arm you. For these reasons your proposal is unacceptable and we are not prepared to discuss it even. (Quoted in Erlich 1994, p. 133)

9. The reunification of Eritrea with Ethiopia and the independence of both Somalia and Sudan meant that Egypt could not bring together what had previously been colonies to form a Nile River Valley union under its control.

10. The emperor of Ethiopia was the country's hereditary ruler until the monarchy was abolished by the military in 1975.

Shortly after this, Nasser's government appeared to have begun a systematic process of undermining and destabilizing the Ethiopian regime. Although there was no public admission by Cairo of a foreign policy directed at such objectives, there is adequate evidence to support the argument that Egypt was so obsessed with guaranteeing its access to the Blue Nile that Cairo was willing to undermine any leadership in Addis Ababa that might threaten that access. One strategy adopted by Cairo was to encourage Ethiopian Muslims to undermine the emperor's government. For example, "Radio Cairo broadcasts started to remind Ethiopian Muslims where their 'primary loyalties' lay" (Kendie 1999, p. 154). The Egyptian government created scholarships for Eritrean Muslim students at Al-Azhar University in Cairo; shortly thereafter, Cairo became an important center for Eritrean students. In 1958 Egypt created a camp near the city of Alexandria for training Eritreans in military tactics. It was at this camp that many of Eritrea's future military leaders obtained their initial training (Kendie 1999, p. 154). Finally, the Egyptian government made available facilities for Eritrean revolutionaries to begin broadcasts that were designed to undermine the Ethiopian government. Many of these broadcasts called on Eritreans "to take up arms and to struggle for their independence" (Kendie 1999, p. 154).

In 1960 Haile Selassie abrogated Eritrea's federal status and turned the territory into Ethiopia's fourteenth province. Shortly after that, the Egyptians established an office in Cairo for what would later become the Eritrean Liberation Front, which became an important leader in the struggle for Eritrean independence. As argued by Daniel Kendie (1999, p. 155), Egypt effectively "succeeded in converting the Eritrean problem into an extension of the Arab-Israeli disputes, and exploited Ethiopia's predicament to its advantage." Of course, though support by Egyptian leaders of the Eritrean struggle may appear to have had opportunistic objectives, many ordinary Egyptians and people in other countries recognized and supported the legitimate claims of Eritreans to self-determination (see Cliffe and Davidson 1988).

Beginning in the early 1950s, there was a significant transformation in the Egyptian political economy. The country began to increase its trade with the Soviet Union. On July 20, 1956, the U.S. secretary of state, John Foster Dulles, withdrew American financial support for the Aswan High Dam, which in turn offered the Soviets an opportunity to influence Egyptian politics.[11] On October

11. In 1956 the United States, the United Kingdom, and the World Bank withdrew their offers to financially support the construction of the Aswan High Dam in their opposition to Egyptian neutrality and the decision by Cairo to align itself with the Soviet Union.

23, 1958, Soviet premier Nikita S. Khrushchev provided Egyptian authorities with a loan of $220 million for preliminary work on the dam (Collins 2002, p. 180), and on August 27, 1960, the Soviets provided Egypt with an additional $120 million for the second stage of the dam construction (Collins 2002, p. 180). Through its work on the Aswan High Dam, the Soviets gained the opportunity to influence not just the Egyptian political economy but also events in the Nile River basin (Kliot 1993). Soviet support actually emboldened the Egyptians in their efforts to exercise hegemony over the region.

The 1929 Anglo-Egyptian treaty was supposed to bind Egypt and the British colonies in the Nile River basin. But Ethiopians have consistently argued that they are not bound by it. They have also questioned the validity of the 1959 bilateral agreement between Egypt and Sudan (Knobelsdorf 2005–06, p. 630).

Ethiopia has been consistent in its dissatisfaction with the Nile Waters agreements. The two most important arguments advanced by Ethiopian authorities against the legal regime represented by the Nile Waters agreements are that it was never a party to either of the two agreements that form this legal regime and that although Ethiopia is the source of the bulk of all Nile River waters, virtually all the Nile River's waters have been allocated to Egypt and Sudan and none to the upstream riparians, including Ethiopia. Hence authorities in Addis Ababa have made statements to the effect that the Nile Waters agreements "are not binding on itself or on other independent basin nations" (Knobelsdorf 2005–06, p. 630; see also Carroll 1999, p. 279). Other upstream riparians have made similar statements, although they have cited different reasons for their decision to declare the Nile Waters agreements null and void and hence not binding on them.[12]

12. For example, see comments by the Kenyan energy minister, Raila Odinga, denouncing the Nile Waters agreements, as reported by Kamau, "Can EA Win the Nile War?," *Nairobi Daily Nation*, March 28, 2002.

5 Theories of Treaty Succession and Modern Nile Governance

On various occasions, many upstream Nile River riparian states, including Ethiopia, Kenya, and Tanzania, have denied the validity of the agreement signed by the downstream riparians—Egypt and Sudan—in 1959 that effectively reserved all of the waters of the Nile River for themselves and left none for the upstream riparians. In their various declarations, these upstream riparian states have sought to reserve the right to claim the waters of the Nile River in the future for their own national development (Kendie 1999, pp. 147–48; Shapland 1997, pp. 74–75). Since they gained independence, many of these states have not developed adequate capacity and the technological know-how to harness the waters of the Nile River and its tributaries for effective local development. Nevertheless, it has been important for them not to surrender their rights to the Nile by acquiescence to agreements that, they strongly believe, were designed to abrogate their rights.

Rapid population increases, coupled with deepening domestic poverty and increased water scarcity, have forced the governments of many of the upstream riparian states to reevaluate their approaches to the management of water in the Nile River basin. Countries such as Kenya, Uganda, Ethiopia, and Tanzania have not only proposed new development projects that would give them access to a significant amount of water but have also become interested in developing irrigation systems to aid their farm sectors, which, since independence, have depended almost exclusively on rain. Some of these countries, notably Kenya, Uganda, and Tanzania, are currently looking at Lake Victoria, an important source of the Nile River, as an anchor to various development projects, while Ethiopia is looking at the Blue Nile for projects to anchor its efforts to significantly improve food security and the supply of reliable and clean sources of energy.

Theories of Succession

When European colonies in Africa began to gain independence in the 1950s and 1960s, there was renewed interest in the continent about the legal effect of changes in sovereignty. The main question was whether the newly independent countries would be bound by treaties entered into on their behalf by their former colonizers. This question was on the minds of the new leaders of Britain's former colonies in the Nile River basin (Uganda, Kenya, Tanzania, and Sudan) who, at independence, had to decide the issue of treaty succession, especially as it related to the Nile Waters agreements.

Some argue that it is the job of international law "to develop some method or device for equitably determining the legal effects of . . . changes of sovereignty" (Wilkinson 1975, p. 22). But what is "state succession"? Is it synonymous with "change in sovereignty"? Daniel Patrick O'Connell defines state succession as "the factual situation which arises when one State is substituted for another in sovereignty over a given territory" (O'Connell 1967, p. 3). Some scholars and practitioners define state succession in terms of competence rather than sovereignty. Hence Sir Humphrey Waldock, who was the United Nations' special rapporteur on the succession of states and governments in respect of treaties in the 1960s, has defined state succession as "a change in the possession of competence to conclude treaties with respect to a given territory" (quoted in Schaffer 1981, p. 593). Waldock, however, after participating in discussions in the International Law Commission of the United Nations about the definition of state succession, adjusted the final definition. In the final draft submitted to the first session of the United Nations Conference on Succession of States in Respect of Treaties, the special rapporteur defined state succession as "the replacement of one State by another in the responsibility for the international relations of territory" (quoted in Schaffer 1981, p. 594).

Scholars of international law suggest that the absence in international customary law of a general rule of state succession to treaties has forced practitioners to look to "competing schools of doctrinal thought" (Schaffer 1981, p. 594). Two theories of state succession have evolved: the theory of universal succession and the so-called clean-slate theory.

Theory of Universal Succession

Universal succession was introduced into international law by Hugo Grotius (1583–1645) (see, for example, Bull, Kingsbury, and Roberts 1992) and is based on analogies extracted from Roman law (see, for example, Wilkinson 1975, p. 432; Schaffer 1981, p. 594; O'Connell 1967, p. 9). Herbert Arnold Wilkinson (1975, p. 13) describes the concept of succession under Roman law as follows: "The 'estate' with its rights and obligations was a legal personality possessed of immortality. Upon the death of the owner, his estate was preserved in its entirety with all its rights and obligations attached." He continues, "Not only did the legal relationship pass from one subject to another, but the subject of that relationship remained the same" (Wilkinson 1975, p. 13). In other words, as Rosalie Schaffer (1981, p. 594) argues, "Under Roman law, rights and duties of the deceased passed *ipso jure* to his successor" (see also O'Connell 1967, p. 9). Since, under Roman law, the rights and duties associated with an estate were derived from absolute natural right, they remained unchanged and retained "their identity despite accidental changes in the identity of those who were from time to time their bearers" (Schaffer 1981, p. 594).

Similarly, a state's personality "was regarded as immutable, irrespective of changes in the identity of its agents, and, as a result neither revolution nor cession of territory could affect the fundamental character and basic constancy of that personality" (Schaffer 1981, p. 594; see also O'Connell 1967, p. 9). Thus once the ruler of a sovereign state had concluded an agreement in the interest of the latter, "his commitments related not only to himself but to the people through whom, in terms of his social contract, he ultimately derived his authority" (Schaffer 1981, p. 594). Rights acquired and obligations incurred by a ruler in the performance of his public duties or functions were expected to continue to bind the state even after the ruler in question had given up his authority (Schaffer 1981, p. 594; see also Wilkinson 1975, pp. 432–34). Under the universal theory of succession, all legal rights and obligations are attached to the land and are considered "'incidental' to the international personality of the state, and consequently, [are] binding on any successor" (Martins 1993, p. 1026).

One can argue that this approach to succession provides a significant level of continuity and stability in the law and effectively eliminates the gap that may exist in the law between the time when the old regime leaves office—either voluntarily or involuntarily—and a new one is installed. It can also be argued that this approach to succession improves efficiency in the economic system: businesses and individuals concluding contracts under one regime need not fear that a succeeding government might abrogate their contractual rights by declaring the old laws to be nonbiding. Nevertheless, over the years, this approach to succession has been considered to be out of step with modern political reality, since it can significantly burden the new state with all the obligations of the *ancien* state. Citizens of the new state may be forced to live with anachronistic and dysfunctional laws and institutions imposed on them by despotic and repressive rulers or with obligations carelessly, opportunistically, and irresponsibly entered into by previous regimes.

This is one of the reasons given by African countries when they state that they are not bound by colonial-era laws and institutions. After all, they argue, most colonial laws and institutions were not designed to enhance peaceful coexistence among Africans nor to provide them with the wherewithal to organize their private lives; they were designed specifically to provide European colonialists with the structures to subject and subdue Africans, exploit the latter and their resources, and generally maximize the objectives of the metropolitan economies and those of the European residents of each colony (see generally Crowder 1987; Fatton 1990; Mbaku 1997, 2004). Thus, many African countries argue, acceptance of the universal theory would be tantamount to an extension of colonial rule.

Clean-Slate Theory

When new states began to emerge in the late nineteenth century, many of them did not want to be bound by universal succession, and hence a new theory developed. The "clean-slate" theory states that "law is an expression of sovereign will and is dependent for its survival on the continued existence of this will" (Schaffer 1981, p. 595; see also O'Connell 1967, pp. 14–17). Thus a change in the sovereignty of a state would mean a collapse of the legal order of the state. That is, the "sovereignty of the predecessor State over the 'lost' territory is abandoned and a legal vacuum is created between the expulsion of one sovereign and the creation of another" (Schaffer 1981, p. 595).

According to the clean-slate theory, then, only the successor state can determine what the nature of the new legal regime will be. Although the successor state may accept and agree to be bound by treaty obligations of the predecessor or extinguished state, this action is purely voluntary and may be based on "considerations of equity, convenience or political interest" and, to a certain extent, political exigency (Schaffer 1981, p. 595). Proponents of the clean-slate theory argue, then, that "the notion of State succession is fallacious and if any continuity of legal rights and obligations does occur in practice, this is entirely attributable to extra-legal factors" (Schaffer 1981, p. 595).[1]

1. The United States is usually considered an early example of a country that adhered to the clean-slate theory. The former British colonies, from which the United States was

Theories of Succession for Modern States

These two succession theories, however, have not been able to meet the demands and expectations of post–World War II independent states, virtually all of which have rejected both theories. The modern approach to state succession to treaties appears to rely more on pragmatism than on strict formalism based on what are considered anachronistic theories of the eighteenth and nineteenth centuries. Although many of the countries that emerged from European colonialism, especially those in Africa, have recognized that some level of reliance on colonial-era agreements is necessary to ensure legal continuity, they nevertheless argue that they cannot be expected to assume responsibility for the treaty obligations of their colonial rulers if they expect to achieve peaceful coexistence and sustainable development in the postindependence period. Part of the problem arises from the fact that many colonial-era agreements, including those in the Nile River basin, were not designed with the participation of the colonized peoples.

Instead, colonial agreements were usually designed to enhance the ability of the reigning European powers to maximize their objectives in the continent—to restructure the colonial territories so that they could serve as sources of raw materials for industrial production in, and markets for the sale of excess output from, the metropolitan economies (see, for example, Lugard 1926; Rudin 1938). Hence overreliance on colonial-era agreements could severely endanger, and in fact derail, postindependence political and economic development (see, for example, Mbaku and Ihonvbere 2003). Perhaps more important is that the new African states believed that they had the right to self-determination and, as sovereign nations, the freedom to determine the kinds and types of laws and institutions that they needed to govern

---

formed, refused to abide by treaties entered into by the British colonial administration. See, for example, Yilma Makonnen (1983, pp. 146–47). In recent years, countries such as Israel and Burkina Faso have relied on the clean-slate theory to denounce treaties entered into by the states that they succeeded. When the colony of Upper Volta gained independence from France, the new country (which later changed its name to Burkina Faso) renounced treaties entered into on behalf of the colony by the French colonial government. Similarly, when the state of Israel came into being in 1948, it refused to be constrained or abide by treaties entered into on behalf of Palestine. See, generally, Beato (1994, especially pp. 539–40 and n. 56). In modern state succession, it is extremely rare to see reliance on either of the two extremes—clean-slate theory or universal succession. States usually approach succession from a pragmatic point of view and make certain that they do not unnecessarily endanger their commercial and other ties to the rest of the world.

their sociopolitical interaction and provide them with the mechanisms for peaceful resolution of conflict, including those arising from trade and other forms of voluntary exchange. Nevertheless, what most African countries now propose is not a return to the clean-slate theory but an approach to treaty succession that grants each former colony (and new sovereign state) the power, freedom, and right to determine which treaties, if any, they may decide to be bound by. This approach has found currency in the Vienna Convention on Succession of States in Respect of Treaties, which states as follows: "A newly independent State is not bound to maintain in force, or to become a party to, any treaty by reason only of the fact that at the date of the succession of States the treaty was in force in respect of the territory to which the succession of States relates."[2]

But what if this results in a treaty vacuum—that is, what if there are no applicable laws to address the particular issues that were the purview of the unilaterally abandoned treaties or laws? Scholars have argued that a treaty vacuum rarely occurs (Schaffer 1981, p. 597). Many newly independent states have created devices they can use to deal with treaties that were entered into on their behalf by their colonial or predecessor states. These mechanisms include, but are not limited to, devolution agreements and unilateral declarations.

DEVOLUTION AGREEMENTS. Many former British colonies in West Africa employed devolution agreements at independence "to secure succession to treaties in general" (Schaffer 1981, p. 597). The devolution agreement was usually concluded between the predecessor state—in the case of the African colonies, the colonial state—and the successor state (that is, the newly independent African state). Generally, each agreement included language to the effect that "treaty rights and obligations of the predecessor State should be assigned or should devolve upon the successor State" (Schaffer 1981, p. 597).

But what exactly is the main objective of a devolution agreement? Usually, the predecessor state enters into such an agreement in an effort to rid or divest itself, and do so in a public manner, of "all rights and responsibilities vis-à-vis the successor State as well as to assign these rights and duties to the latter" (Schaffer 1981, p. 599). If the devolution agreement is registered with the United Nations and is then published by that world body, that will put third states and other nonparticipating parties on notice about the devolution. A. P. Lester (1963, p. 504) argues, however, that when these agreements are registered with the secretary-general of the United Nations, they are "regarded

2. Vienna Convention on Succession of States in Respect of Treaties, Vienna, August 23, 1978; U.N.T.S. 1946, art. 16.

as binding in international law." Rosalie Schaffer (1981), however, doubts whether mere registration with the United Nations alone would "today be sufficient to render a devolution agreement absolutely binding on the parties, especially in view of Article 8(1) of the Convention on Succession of States in respect of Treaties." Article 8(1) states as follows:

> The obligations or rights of a predecessor State under treaties in force in respect of a territory at the date of a succession of States do not become the obligations or rights of the successor State towards other States Parties to those treaties by reason only of the fact that the predecessor State and the successor State have concluded an agreement providing that such obligations or rights shall devolve upon the successor State.[3]

Taking this article into consideration, one can argue that each successor state must commit some other overt act (besides registration with the United Nations) to indicate to the world its intention to abide and be bound by the terms of the devolution agreement and consequently to succeed to treaties entered on its behalf by the predecessor state. Kenneth J. Keith (1967, pp. 540–41) argues that even though devolution agreements

> do not bind third parties, they have considerable effect: (a) they demonstrate that many newly independent states consider they remain bound by at least some treaties; the parent state also must have this view; (b) they have actual contractual effect between the parties; and not only between the parties to each specific agreement but also between all the parties to all of the agreements; (c) they inevitably give rise to estoppels, especially as they are given content by the practice such as that reviewed above; (d) they may assist the parent state to rid itself of any obligations in respect of its former dependent territories.[4]

3. Ibid., art. 8(1).

4. Footnotes present in the original quotation are not included. Estoppel is a legal principle designed to prevent, for example, an individual from denying a settled fact. According to *Black's Law Dictionary,* estoppel is "a bar or impediment raised by law, which precludes a man from alleging or from denying a certain fact or state of facts, in consequence of his previous allegation or denial or conduct or admission, or in consequence of a final adjudication of the matter in a court of law." For example, a creditor promises to accept a lesser sum of money in full payment for the debt, intending the debtor to rely on that promise. The debtor does rely on that promise. If the creditor later sues for the balance of the debt, the debtor may have a defense of promissory estoppel.

But what about any possible impacts or effects on third parties of devolution agreements? The general rule is that third states, which either are not signatories to the devolution agreement or have not acquiesced to it, are not bound by the agreement's terms (see, for example, Schaffer 1981, p. 600). All devolution agreements are usually placed with the secretary-general of the United Nations; hence third states would normally be formally informed or given notice of "the intended assignment of treaty rights and obligations" (Schaffer 1981, p. 600). Schaffer argues that if "the third State acknowledges this assignment, the devolution agreement will most probably result in a novation of treaties. If it does not formally acknowledge the assignment but nevertheless continues to exercise rights and to discharge obligations under the treaty, a novation may be implied."[5] But what if a third state simply does not act? As O'Connell (1967, p. 372) argues, in the case of incompatible reservations to a treaty, as well as devolution agreements, third states that stay silent are considered to have tacitly consented (see also Schaffer 1981, p. 601).

Article 8 of the United Nations Convention on Succession of States in Respect of Treaties deals with "devolution of treaty obligations or rights from predecessor State to a successor State," but it does not address either explicitly or implicitly the twin issues of express and tacit consent by third states to devolution agreements. However, section 2 of that article states that "notwithstanding the conclusion of such an agreement, the effects of a succession of States on treaties which, at the date of that succession of States, were in force in respect of the territory in question are governed by the present Convention." Schaffer (1981, p. 601) argues that the "omission [of any provisions dealing with the relationship between third States and successor States] was deliberate and aimed to avoid the difficult questions raised of the rights and obligations of third States arising out of a devolution agreement." Article 8 simply makes it clear that successor states are not bound "in a treaty relationship with third States" (Schaffer 1981, p. 601).

UNILATERAL DECLARATIONS. Many independent African states, especially those in East and Central Africa, have opted to rely on unilateral declarations as a way to deal with issues of treaty succession. Under this approach, the new country declares that it will temporarily maintain the inherited treaty obligations to give it time to determine which treaties it would permanently abandon and which it would retain and make part of its permanent legal framework.

5. Schaeffer (1981, p. 600). *Novation* is a legal term that means the substitution of a new debt or obligation for an existing one.

Julius Kambarage Nyerere became the prime minister of an independent Tanganyika in December 1961. Tanganyika later merged with the islands of Zanzibar in 1964 to form the United Republic of Tanzania. In 1961 Nyerere proposed what came to be known as the Nyerere doctrine of treaty succession (see, for example, Kiplagat 1995, pp. 263–64; Schaffer 1981, p. 602; Knobelsdorf 2005–06, p. 622). In a letter dated December 9, 1961, and sent to the secretary-general of the United Nations, the government of Tanganyika made clear its intention to be bound, at least temporarily, by bilateral treaties "validly concluded by the United Kingdom on behalf of the territory of Tanganyika." Prime Minister Nyerere's letter read as follows:

> The Government of Tanganyika is mindful of the desirability of maintaining, to the fullest extent compatible with the emergence into full independence of the State of Tanganyika, legal continuity between Tanganyika and the several States with which, through the action of the United Kingdom, the territory of Tanganyika was prior to independence in treaty relations. Accordingly, the Government of Tanganyika takes the present opportunity of making the following declaration:
>
> As regards bilateral treaties validly concluded by the United Kingdom on behalf of the territory of Tanganyika or validly applied or extended by the former to the territory of the latter, the Government of Tanganyika is willing to continue to apply within its territory, on a basis of reciprocity, the terms of all such treatments for a period of two years from the date of independence (that is, until December 8, 1963) unless abrogated or modified earlier by mutual consent. At the expiry of that period, the Government of Tanganyika will regard such of these treaties which could not by the application of the rules of customary international law be regarded as otherwise surviving, as having terminated.
>
> The Government of Tanganyika is conscious that the above declaration applicable to bilateral treaties cannot with equal facility be applied to multilateral treaties. As regards these, therefore, the Government of Tanganyika proposes to review each of them individually and to indicate to the depositary in each case what steps it wishes to take in relation to each such instrument—whether by way of confirmation or termination, confirmation of succession or accession. During such interim period of review any party to a multilateral treaty which has prior to independence been applied or extended to Tanganyika may, on a basis of reciprocity, rely as against Tanganyika on the terms of such treaty.[6]

6. The text of the Nyerere letter to the secretary-general of the United Nations regarding treaty succession can be found in *United Nations Conference on Succession of States in*

As Tanganyika argued in its unilateral declaration, all personal treaties would lapse after the two-year period, but the country would succeed to the real treaties (Schaffer 1981, p. 603), those that dealt with boundary and communication.[7] This was in line with the clean-slate theory, which "presupposes the total destruction of the previous State identity and the creation of a new international legal person" (Maluwa 1999, p. 70) and "attributes to the new state a basic right to legal self-determination" (Knobelsdorf 2005–06, p. 633).

On July 2, 1962, Britain's permanent representative to the United Nations, following Tanganyika's unilateral declaration, "sent a disclaimer of responsibility to the [UN] Secretary-General which was circulated among member States" (Schaffer 1981, p. 603). In the disclaimer, the British government stated that as from the date at which Tanganyika gained independence and became a sovereign state, the United Kingdom had "ceased to have the obligations or rights which they formerly had as the authority responsible for the administration of Tanganyika."[8]

Uganda gained independence from Great Britain in 1962 and elected not to sign a devolution agreement with its former colonizer. Instead, Uganda's new leaders chose to follow the example of Tanganyika and on February 12, 1963, sent a letter to the secretary-general of the United Nations, which was subsequently circulated to member states.[9] Uganda's unilateral declaration was, in substance, similar to that advanced by the government of Tanganyika. Uganda, however, limited its period of review to fourteen months, in respect of both multilateral and bilateral treaties (Tanganyika had chosen a review period of two years). Following Uganda's declaration, Britain's representative to the United Nations sent a disclaimer of responsibility to the secretary-general, which was subsequently circulated among member states.[10]

Other declarations, similar to those issued by Tanganyika and Uganda, were made by the governments of Botswana, Lesotho, Kenya, and Malawi (Schaffer 1981, p. 604). The basic theme running through virtually all of these

---

Respect of Treaties, 1977 Session and Resumed Session 1978, Vienna, 4 April–6 May 1977 and 31 July–23 August 1978, Official Records III, A/CONF.80/16/Add.2 (http://untreaty.un.org/cod/diplomaticconferences/succ-treaties-1978/vol/english/vol_III_e.pdf).

7. For an in-depth discussion of real and personal treaties, as these terms relate to the Nyerere doctrine generally and the Tanganyika parliamentary debates that preceded Nyerere's declaration, see O'Connell (1967).

8. *Materials on Succession of States*, U.N. Doc. ST/LEG/SER.B/14(1967), p. 178.

9. The text of the letter sent by the government of Uganda can be found in ibid., pp. 179–80.

10. Ibid.

declarations was that the newly independent countries did not want to be permanently bound by treaties that they did not take part in negotiating or creating and about which they were not consulted.

Succession Theories and Nile Governance

The independence of the European colonies in Africa involved a change of sovereignty that created complex legal problems; the colonial state ceased to rule the territory and the postcolonial state emerged to take over. As argued by D. P. O'Connell (1967, p. 3), such a change produced significant legal consequences for the territory's "economic, social and legal structure." A body of law whose main objective is to deal with the problems arising from this "transfer of territory" is called the "law of State succession" (O'Connell 1967, p. 3). Here, we are particularly interested in the Nyerere doctrine of state succession and its relevance to governance in the Nile River basin.

Nyerere Doctrine

Followers of the Nyerere doctrine argue that the Nile Waters agreements place the upstream riparian states at the mercy of Egypt and force them to subject their development plans to the scrutiny and supervision of Egyptian authorities, and that such an approach to development is not compatible with their status as sovereign independent states, each with its own international legal identity (see, for example, Knobelsdorf 2005–06, p. 631). This, essentially, is the argument being advanced by the upstream riparian states, all of which did not participate in the creation of the Nile Waters agreements.[11]

In 1962, following its unilateral declaration, the government of Tanganyika sent a letter to the government of Egypt regarding the Nile Waters agreements and stated that "an agreement purporting to bind [upstream riparians] in perpetuity to secure Egyptian consent before undertaking its own development programs based on its own resources was considered to be incompatible with Tanganyika's status as a sovereign state" (quoted in Knobelsdorf 2005–06, p. 632). Tanganyika's decision to take this approach to the governance of the Nile River was based on its belief that "the colonial treaties are not binding on the

11. One can argue that this is an exaggeration of the actual situation in the Nile River basin, given that the upstream riparian states can simply ignore the Nile Waters agreements and proceed with their development plans. Should this happen, Egypt would be faced with taking either legal or military action to arrest the situation. Of course, Egyptians would have to weigh the short- and long-term costs and benefits of each line of action.

newly independent states because the new states never took part in the negotiations creating the obligations under these treaties" (Carroll 1999, p. 279).

From the point of view of international law, Tanzania and any other country that repudiated succession treaties immediately after independence may have significantly more legal ground to stand on than those countries that are only now engaging in renunciation.[12] In December 2003, the government of Kenya made a declaration similar to those that had been issued by Tanzania and Uganda following their independence. The Kenyan Parliament stated that Kenya, which had not been a party to the 1929 Anglo-Egyptian treaty and had not been consulted before the protocol was enacted, would not consider the agreement binding. Kenyan authorities regretted having not followed the example of Tanzania and made a unilateral declaration, and hence repudiating the Nile Waters agreements, at independence. However, many of these leaders felt that it was time to repudiate what they believed were colonial-era impositions that continue to significantly constrain the government of Kenya's ability to effectively use the country's natural resources, including its water resources, to deal with domestic poverty and improve the living conditions of the country's citizens (Knobelsdorf 2005–06, pp. 633–34).

Ethiopia as a Unique Upstream Riparian

Like other upstream riparian states, Ethiopia has also repudiated the Nile Waters agreements, arguing that it was never a party to either of the Nile Waters agreements nor was it consulted before any of the protocols were enacted. More important is Ethiopia's argument that since it was never subject to long-term colonization, unlike the other riparian states, no colonial government ever negotiated a treaty on its behalf. Thus as a sovereign independent country, which was not consulted, none of the treaties supposedly governing the Nile River is binding on Ethiopia and its peoples (see, for example, Carroll 1999, p. 279). Ethiopia, however, has not followed a modified clean-slate doctrine of treaty succession. Instead, it has taken what has been referred to as a "developmental approach," wherein former colonies denounce or repudiate treaties that were negotiated on their behalf by their former colonizers but have become anachronistic or no longer meet their developmental needs.[13] The general argument is that a country should not be forced to adhere

12. Tanganyika united with the islands of Zanzibar in 1964 to form the United Republic of Tanganyika and Zanzibar. Later that year, the country was renamed United Republic of Tanzania.

13. Mention has also been made of the Harmon doctrine, which "holds that a country is absolutely sovereign over the portion of an international watercourse within its

to an agreement that is outdated and no longer serves the domestic public interest, especially if that agreement was designed without the participation of the society that it now governs.[14] Like the citizens of most of the other riparian states, Ethiopians believe that an appropriate legal framework for the governance of the Nile River basin must be one that reflects the interests and aspirations, especially for economic growth and development, of the relevant stakeholders and provides the wherewithal for the people to deal effectively with ever-evolving problems (for example, global warming, droughts, rapid population growth, industrialization, migration and other shifts in population, and globalization) on the ground as the rules are implemented (Carroll 1999, p. 279).

Of course, Ethiopia did sign an agreement with Britain in 1902 (the 1902 Anglo-Ethiopian agreement). Nevertheless, shortly after he was returned to power after the Italians were ousted from the country in 1941, the Ethiopian leader Haile Selassie repudiated the 1902 treaty in retaliation for British support of Italy during the latter's occupation of Ethiopia from 1936 to 1941 (Kendie 1999, p. 147). Additionally, Emperor Menelik II, who signed the agreement on behalf of Ethiopia, is said to have done so as a result of a "mistranslation between the English and Amharic version" (Lie 2010, p. 7). The English version of article 3 of the 1902 treaty read partly as follows: Emperor Menelik II was "not to construct or allow to be constructed any work across the Blue Nile, Lake Tsana or the Sobat, which would *arrest* the flow of their waters into the Nile except in agreement with His Britannic Majesty's Government and the Government of Soudan."[15] In the Amharic version, the word *arrest* was translated as "stop," so that the emperor understood the agreement

---

borders. Thus that country would be free to divert all of the water from an international watercourse, leaving none for downstream states" (McCaffrey 1996, p. 549).

14. Egypt's argument with respect to Ethiopia and the allocation of the waters of the Nile River rests primarily on what Egyptian leaders believe are rights to the use of the river's waters—"natural and historical rights" (Mekonnen 2010, p. 432)—that were won through actual harvest and use of the waters of the Nile River over the years and that, moreover, these rights were formally elaborated in agreements signed in 1929 and 1959. However, though the Nile Waters agreements may grant rights to the use of the waters of the Nile River to Egypt and the Republic of Sudan, these rights are relevant only to actions between the states (that is, Egypt and the Republic of Sudan) that are legally bound by these treaties. This is not so with regard to the upstream riparian states—that is, the parties that are not bound by the 1929 and 1959 treaties.

15. Protocol between the Governments of Great Britain and Italy, for the Demarcation of Their Respective Spheres of Influence in East Africa, U.N. Legislative Series 1963, p. 112, quoted in Ullendorff (1967, p. 643). Emphasis added.

as saying that as long as he or his activities on the Blue Nile did not stop the flow of water into Egypt, he was free to divert and use waters of the Blue Nile.

In conclusion, then, the Ethiopians argue that they are not bound by the Nile Waters agreements because they were not signatories to any of these agreements. Kenya, Uganda, and Tanzania have argued that while some level of reliance on treaties signed on their behalf by their colonial rulers is necessary to ensure legal continuity, they would reject those colonial-era treaties, such as the Nile Waters agreements, that they believed would harm their national interests.

6 International Water Law and the Nile River Basin

The management of an international watercourse, such as the Nile River, requires cooperation among all the watercourse's riparian states. Such cooperation should deal specifically with allocation and use of the watercourse's resources, including its water. In addition, riparians should formulate basic standards and rules to cooperate on pollution, overexploitation, and ecosystem degradation. An international watercourse convention can provide the foundation for the necessary cooperation between all of the watercourse's riparian states.

The UN Watercourses Convention

Examining laws governing or regulating international watercourses can provide insights into how to deal with the Nile River, which itself is an international watercourse. An important international watercourse law is the UN Watercourses Convention.[1] The Nile River riparian states and other stake-

1. UN Convention on the Law of the Nonnavigational Uses of International Watercourses, May 21, 1997, G.A. Res. 5/229, U.N. GAOR, 51st sess., 99th plen. mtg., U.N. Doc. A/RES/51/229 (1997) (https://treaties.un.org/Pages/ViewDetails.aspx?src=IND&mtdsg_no=XXVII-12&chapter=27&lang=en). According to the convention's article 36(1), "The present Convention shall enter into force on the ninetieth day following the date of deposit of the thirty-fifth instrument of ratification, approval or accession with the Secretary-General of the United Nations." On May 21, 2014, Vietnam became the thirty-fifth contracting state to the convention, making it possible for the latter to enter into force, on August 17, 2014.

holders of the Nile may find an examination of the convention informative and beneficial. Of especial interest to Nile River stakeholders, including the basin's policymakers, is article 4(1), which states that "every watercourse State is entitled to participate in the negotiation of and to become a party to any watercourse agreement that applies to the entire watercourse, as well as to participate in any relevant consultation." Given that none of the upstream riparian states participated in any of the negotiations that produced the Nile Waters agreements, one can argue that based on international law principles, the existing legal framework in the Nile River basin, as embodied in the Nile Waters agreements, should be abandoned in favor of participatory and inclusive agreements to produce a more effective legal compact.

The "equitable and reasonable utilization and participation" principle articulated in article 5(2) is usually considered the heart of the UN Watercourses Convention, and it is the principle that is most likely to be of interest to the Nile River's upstream riparians. Article 5(2) states that "Watercourse States shall participate in the use, development and protection of an international watercourse in an equitable and reasonable manner." Article 6 provides mechanisms to determine what constitutes reasonable and equitable use: to make such a determination, certain factors need to be considered, including but not limited to geography; hydrology; climatic conditions; past, present, and potential water uses; population; economic and social needs of each basin state; comparative costs of alternative means of meeting the economic and social needs of each basin state; availability of other resources; and cost minimization in the utilization of the waters of the basin.

Another foundational principle that is considered critical to the management of international watercourses is the "obligation not to cause significant harm." Article 7(1) of the convention states that, in exploiting the waters of an international watercourse, for example, for domestic and commercial purposes, states are required to "take all appropriate measures to prevent the causing of significant harm to other watercourse States." Articles 5, 6, and 7 reinforce each other to provide mechanisms for all watercourse riparian states, both upstream and downstream, to deal fully and effectively with conflicts arising from the use of the watercourse.

Stephen C. McCaffrey, who served as chair of the International Law Committee Working Group on the Draft Articles of the UN Convention on International Watercourses, has offered the following interpretation of how these provisions should work:

> If a State believes it has sustained significant harm due to a co-riparian State's use of an international watercourse, it will ordinarily raise the

> issue with the second State. In the negotiations that follow, articles 5, 6, and 7 in effect provide that the objective is to reach a solution that is equitable and reasonable with regard to both States' uses of the watercourse and the benefits they derive from it. The possibility that the solution may include the payment of compensation, to achieve an equitable balance of uses and benefits, is not excluded.[2]

All watercourse states are obligated to cooperate as demanded by article 8 of the convention, and they are expected to do so "on the basis of sovereign equality, territorial integrity, mutual benefit and good faith in order to obtain optimal utilization and adequate protection of an international watercourse" (art. 8[1]). Article 9 provides for the "regular exchange of data and information" on water quality and on the "hydrological, meteorological, hydrogeological and ecological nature" of the watercourse in question (art. 9[1]).

Articles 11 to 19 of the UN Watercourses Convention is devoted to "planned measures"—that is, the obligation on the part of watercourse states to notify any parties that might be affected by any projects that are planned on the watercourse. To fulfill this obligation, states must make "timely notification" and provide "available technical data and information, including the results of any environmental impact assessment" (art. 12). The state that has been notified is expected to reply quickly but is granted up to six months to comply (arts. 13 and 15).[3] If the watercourse state that has been notified believes that the planned or impending project would be inconsistent with or violate the provisions of either article 5 or 7, that state must undertake negotiations to secure an equitable resolution of the conflicting issues.[4]

In articles 20–26, the convention sets out the legal mechanisms for jointly protecting, preserving, and managing international watercourses. This part of the convention provides specifically for the "prevention, reduction, and control of pollution" (art. 21) and requires all watercourse states to take all necessary and appropriate steps or measures to prevent and mitigate any conditions related to the international watercourse that might be harmful to other watercourse states (art. 27).

2. S. C. McCaffrey, "Convention on the Law of Nonnavigational Uses of International Watercourses," UN Audiovisual Library of International Law, 2014 (http://legal.un.org/avl/ha/clnuiw/clnuiw.html).

3. See also Carroll (1999).

4. Here, the states affected include both the "notifying" states and the "notified" states. UN Watercourses Convention, art. 17(1).

Lessons for the Nile River Basin

According to article 3(1) of the UN Watercourses Convention, "In the absence of an agreement to the contrary, nothing in the present Convention shall affect the rights or obligations of a watercourse State arising from agreements in force for it on the date on which it became a party to the present Convention." In other words, the UN Watercourses Convention "does not affect the status or obligations of existing agreements" (Carroll 1999, p. 286), such as the 1929 Anglo-Egyptian treaty and the 1959 bilateral agreement between Egypt and Sudan, in the case of the Nile River basin.[5] Parties to an international watercourse, such as the Nile River, can jointly negotiate, compact, and adopt a new regional legal framework that can coexist with the UN Watercourses Convention (Carroll 1999, p. 286). Given that, especially according to the upstream riparian states, the Nile River basin needs a new, more inclusive, and regionally focused legal framework to deal fully and effectively with "current environmental and water allocation issues" (Carroll 1999, p. 286), the eleven riparian states of the Nile River basin can make use of the UN Watercourses Convention as the "basis for a new Nile agreement" (Carroll 1999, p. 286).[6]

However, any effort to employ the UN Watercourses Convention as a foundation for the development of a Nile River legal framework could confront many problems (Carroll 1999, pp. 287–88). First, many of the Nile River basin riparian states either voted against the UN Watercourses Convention or abstained from voting (Carroll 1999, p. 287).[7] Second, none of the eleven Nile River basin states have signed or ratified the UN Watercourses Convention since it was adopted in 1997.[8] At the level of the working group, several of the Nile River riparian states raised objections that betrayed their unwillingness to employ the convention as a guide or foundation for the development of a new Nile River legal framework (Carroll 1990, p. 287). For example, Ethiopia,

5. Exchange of Notes between His Majesty's Government in the United Kingdom and the Egyptian Government in Regard to the Use of the Waters of the River Nile for Irrigation Purposes (with Seven Diagrams), Cairo, May 7, 1929, L.N.T.S. 2103; United Arab Republic and Sudan Agreement (with Annexes) for the Full Utilization of the Nile Waters, Cairo, 8 November 1959, 6519 U.N.T.S. 63.

6. See article 3 of the UN Watercourses Convention, which allows for the development of legal frameworks that coexist with the convention.

7. Egypt, Ethiopia, Rwanda, and Tanzania abstained from voting. Democratic Republic of Congo and Uganda did not take part in the vote. Burundi voted against, and only Kenya and Sudan voted in favor.

8. UN Watercourses Convention.

Rwanda, and the Republic of Sudan disagreed with article 32's nondiscrimination clause.[9]

Third, the principle of "equitable and reasonable utilization" presented in article 5 is vague and hard to apply. According to Christina M. Carroll (1999, p. 287), the convention allows each watercourse state to take into consideration a litany of "factors and circumstances" (art. 6[1][a–g]), all of which "have no given weight, and thus it may be difficult to reach agreement on what combination of factors constitutes equal utilization" (Carroll 1999, pp. 287–88).

Principle of Equitable and Reasonable Utilization

Significant or even slight variations in water needs, and to a greater extent, demands, especially when compared with the available supply of water, could subject article 6's principle of equitable and reasonable utilization to the types of interpretations that can make reaching agreement difficult or virtually impossible. Any attempt to apply article 6's principle to the development of a legal framework for Nile River governance must consider the problems associated with interpretation.

Egypt

Egypt suffers from what many scholars have termed the "tyranny of dependency" on the waters of the Nile (Collins 2002, p. 11). Consider the following words from an old hymn: "O, Nile, verdant art thou, who makes man and cattle to live."[10] Presently, Egypt relies wholly on the waters of the Nile River for all its commercial and domestic water needs. It is no wonder that Egyptian authorities, over the years, have made a claim to what they call "natural and historical rights" (Mekonnen 2010, p. 432; see also art. 1 of the 1959 Agreement between Egypt and Sudan). Some Egyptians have argued that these rights derive from the fact that Egypt has used the waters of the Nile River for millennia without any credible challenge from any of the upstream states. This, of course, is not true: Ethiopia has challenged and raised objections to

9. Ibid. Article 32 reads as follows: "Unless the watercourse States concerned have agreed otherwise for the protection of the interests of persons, natural or juridical, who have suffered or are under a serious threat of suffering significant transboundary harm as a result of activities related to an international watercourse, a watercourse State shall not discriminate on the basis of nationality or residence or place where the injury occurred, in granting to such persons, in accordance with its legal system, access to judicial or other procedures, or a right to claim compensation or other relief in respect of significant harm caused by such activities carried on in its territory."

10. *Hymn to the Nile, Papyrus Sallier II*, quoted in Collins (2002, p. 11).

Egyptian interference in its use of the Blue Nile, the most important tributary of the Nile River. Other Egyptian leaders have based Egypt's claims to the waters of the Nile River on what they believe were property rights assigned to them by treaties signed in 1929 and 1959. However, property rights are relative: though treaties such as the 1929 Anglo-Egyptian treaty and the 1959 bilateral agreement between Egypt and Sudan may "grant" rights between those states that are actually bound by the treaties, they cannot do so relative to parties that are not bound by the treaties.

Despite the fact that the upstream riparian states have expressly denounced the Nile Waters agreements, and that none of the upstream riparian states is a party to any of the agreements, Egypt continues to insist that the rights "granted" by the 1929 and 1959 treaties are valid as between all the Nile River's riparian states. Since the Nile Basin Initiative went into force, Egyptian authorities have made relatively aggressive efforts to prevent the development and adoption of any legal framework that, in their opinion, would threaten the rights granted them by the Nile Waters agreements.[11]

Stephen C. McCaffrey and Mpazi Sinjela (1998, p. 100) report that during the working group negotiations leading to the compacting of the UN Watercourses Convention, representatives of Egypt argued that "the availability of other water resources" should be taken into consideration when making a determination about what is equitable utilization under the convention's article 6. Although the working group did not accept Egypt's suggestions, it nevertheless added, as a factor to be considered in determining equitable and reasonable utilization, "the availability of alternatives, of comparable value, to a particular planned or existing use."[12]

11. See, for example, Mekonnen (2010, pp. 421–40). According to the Nile Basin Initiative's website, "The Nile Basin Initiative (NBI) is a regional inter-governmental partnership led by 10 Nile riparian countries, namely Burundi, Democratic Republic of Congo, Egypt, Ethiopia, Kenya, Rwanda, South Sudan, Sudan, Tanzania and Uganda." Because Eritrea opted not to become a member of NBI, it can only participate as an observer. The initiative "provides riparian countries with the only all-inclusive regional platform for multistakeholder dialogue, information sharing as well as joint planning and management of water and related resources in the Nile Basin" ("About the Nile Basin Initiative" [www.nilebasin.org]). In recent years, NBI countries, particularly the upstream riparian states, have made concerted efforts to develop an alternative legal framework to the Nile Waters agreements. See, for example, Agreement on the Nile River Basin Cooperative Framework (www.internationalwaterlaw.org/documents/regionaldocs/Nile_River_Basin_Cooperative_Framework_2010.pdf).

12. UN Watercourses Convention, art. 6(g). The working group, however, did not make clear whether in considering those alternative water resources, one needs to be aware or cognizant of such variables as the quality of the water, the reliability the source,

For many years, Egyptians have argued that without the Nile River, there is no Egypt or Egyptian civilization. They argue that their very existence depends on the waters of the Nile River (see, for example, Kendie 1999). For millennia, Egyptians have used the waters of the Nile River for commercial (for example, agriculture) and domestic purposes, and during this period, the country has formed a unique relationship with the river. Thus some of today's Egyptians have argued that since they were the first civilization to put the Nile River's waters to beneficial use, and given that their very survival depends almost exclusively on the waters of the Nile River, their present use must be considered reasonable and equitable (Carroll 1999, p. 288). Additionally, the Egyptians have argued that once "the population dependent on the watercourse" provision of article 6(1)(c) is taken into consideration, Egypt must be allowed to maintain the historical water allocation that its population has depended on for millennia.[13]

Ethiopia

Ethiopians view equity quite differently. For one thing, Ethiopians are likely to argue that since the country's highlands provide more than 80 percent of the waters that flow into the Nile River, any treaty that regulates access to the waters of the Nile River should grant them, at the minimum, an annual share that is greater than that currently granted to Egypt (Brunnée and Toope 2002, p. 117).[14] Given that contribution of water to the Nile River is not one of the relevant factors enumerated in article 6 for determining equitable and reasonable utilization, this argument is not sustainable under the convention.[15]

Ethiopia, of course, can invoke other article 6 factors or provisions to support its argument that equity requires that it be allocated more of the waters of the Nile River. Ethiopian authorities could support their argument for more water by invoking the country's "social and economic needs"; the

---

and the ability of the country involved to reasonably access the water, given existing technology and cost constraints.

13. See the 1959 bilateral agreement between Egypt and Sudan, art. 2(4). As much as 96 percent of the population of Egypt lives on the banks of and depends entirely on the Nile River and the its delta. See, for example, Kendie (1999, p. 142).

14. Specifically, three rivers—the Blue Nile (which drains Lake Tana), the Atbara, and the Sobat—provide the bulk of the water that flows into the Nile River. The rest of the Nile River's waters come from the White Nile, which originates in Lake Victoria.

15. Article 6 of the UN Watercourses Convention provides "relevant factors and circumstances" that must be used to determine whether an international watercourse is being utilized equitably and reasonably. See also Carroll (1999, pp. 288–89).

dependence of its population on the watercourse; lack of "availability of alternatives, of comparable value, to a particular planned, or existing use; and the significant impact that Egypt's current use could have on Ethiopia's ability to harvest and use the waters for its own development (art. 6[b, c, g, d]). It appears, then, that both Ethiopian and Egyptian officials could use article 6 in their arguments and come to completely different outcomes regarding equitable and reasonable use of the waters of the Nile River.

Under the convention's article 6(g), the water uses of other riparian states must be considered when equitable use is determined. However, the term *comparable use* is not expressly defined, and its use is subject to interpretation. In fact, the government of each riparian state could produce an interpretation of comparable use that would fully support its own concept or view of equitable and reasonable utilization (Carroll 1999, p. 289).

Preventing Significant Harm

Article 7 of the UN Watercourses Convention states that "in utilizing an international watercourse in their territories," all riparian states are required to "take all appropriate measures to prevent the causing of significant harm to other watercourse States" (art. 7[1]). What exactly are appropriate measures, and, as argued by some scholars (for example, Carroll 1999, p. 289), what activities or actions would constitute the duty to prevent or minimize damage to the interests of other users of the watercourse?

Because the word *harm* is not defined by the convention, disputing or affected parties would have to determine, first, what constitutes harm, and second, when it is actually imposed on one state by another. What, then, is "significant harm"? Would the construction of the Grand Ethiopian Renaissance Dam on the Blue Nile, for example, constitute significant harm to Egyptian interests under the convention? More important, is harm to another limited to activities that reduce the flow of water to others, or does it include those activities that degrade the quality of the watercourse? Thus, since the UN Watercourses Convention does not provide a definition of *harm*, the application of this principle is likely to be quite difficult.

If the Nile River basin countries employ the do-no-harm principle to design a new legal framework or incorporate it and other of the convention's principles into such an agreement, the conflict over "equitable utilization" would not necessarily be resolved. According to Christina M. Carroll (1999, p. 290), Ethiopia has argued that the no-harm principle should only be invoked and made operational when a watercourse state "has exceeded its equitable or reasonable use." When one takes into consideration the positions taken by both the upstream and downstream riparian states regarding use of the waters of the Nile River, incorporating the principles of equitable and reasonable utilization and the obligation not to cause significant harm into a new Nile River basin legal framework would essentially "pit upstream and downstream states against each other" (Carroll 1999, p. 290).

The Nile Basin Initiative and the Cooperative Framework Agreement

In 1999 the Nile River basin countries, working within the Nile Basin Initiative, forged ahead with negotiations to develop a new legal framework for the Nile River, one that they hoped and believed would replace the Nile Waters agreements and generally strengthen cooperation among riparian states, especially with respect to the management of the waters of the Nile River.[16] The proposed legal framework, officially known as the Nile River Basin Cooperative Framework Agreement (CFA), represented the first effort by the Nile River riparians to introduce the principle of equitable and reasonable water allocation and utilization into any talk about governance of the Nile River.

The designers of the cooperative framework sought to replace unilateralism and competitive nationalism with a regional and cooperative approach to resolving and dealing with Nile River issues. Arguing that the Nile Waters agreements were the most important constraint to cooperation in the basin, the participants in this exercise believed that a new, regionally based cooperative agreement, which could replace colonial-era treaties, should be created. Unfortunately, the development and adoption of the framework has stalled because of disagreements between the upstream and downstream riparian states over one of the Cooperative Framework Agreement's provisions—article 14(b). Egypt and the Republic of Sudan, the downstream riparians, have insisted that the wording of that article be changed to guarantee that they retain the water rights granted them by the Nile Waters agreements. The upstream riparian states argue that only rights granted by the cooperative framework should be respected and that it is only through such a new legal

16. The Nile basin countries are Ethiopia, Sudan, Egypt, Tanzania, Eritrea, Kenya, Uganda, Democratic Republic of Congo (DRC), Burundi, and Rwanda. Shortly after it became independent in 2011, South Sudan became the eleventh riparian state. Both the Nile Basin Initiative and the new legal framework, called the Cooperative Framework Agreement, are examined in more detail later.

framework that the basin will achieve equity and reasonableness in the extraction and use of the Nile River's waters.

Legal Systems for Water Allocation

Although they have not expressly done so, the Egyptians seem to be relying on what the literature on water law refers to as the riparian and prior-appropriation doctrines (Getches 1990, pp. 3–8). Egyptians have, over the years, claimed a natural right to the use of the waters of the Nile River because the river runs through their land mass—in other words, the country is a riparian state. They have also claimed an acquired right granted to them by the Nile Waters agreements and based on the fact that they were among the earliest modern humans to put the waters of the Nile River to beneficial use. Two important legal systems for water allocation deal with the basis of Egypt's claims.

Riparian Rights

Under the riparian doctrine, individuals whose land borders a waterway are granted certain appurtenant rights. These rights allow the landowners to harvest and utilize reasonable amounts of water, as well as access the water for other purposes, including, for example, hunting, fishing, boating, and recreation (Getches 1990, p. 4). In some countries that subscribed to this doctrine, there evolved a rule call natural flow, under which each riparian owner was granted "the right to have water flow past the land undiminished in quantity or quality" (Getches 1990, p. 4). In the United States, the courts used case law to temper perceived harshness in the doctrine with "reasonable use." Today, in the United States, the doctrine of riparian rights has evolved significantly, and reasonable use now represents a critical element of its application.

First, where there is not enough water to meet the reasonable needs of all riparians, all of them must reduce their water use in proportion to their rights. Second, the right to use water is generally not extended to nonriparians (that is, owners of nonriparian land), although nonriparian land may use water if it compensates riparians for the harm caused them by such use. Third, riparian rights are inherent in ownership of the land that is riparian (to a watercourse) and, as a result, the owner of these rights does not need to regularly exercise them in order for them to remain valid. Thus a riparian landowner who has not used his rights for some time may resume use, forcing other users to make necessary adjustments to accommodate such newly initiated use. Fourth, riparian rules in the United States and other countries have, over the years, been altered and modified by legislative acts and case law. Today, in many countries with riparian traditions, riparian landowners must apply for and obtain permits from the government in order to use water. Fifth, in these countries, nonriparians can also secure such permits to use water. Finally, riparian landowners still retain special rights to use "the surface of waters adjoining their property" (Getches 1990, p. 6). Nevertheless, if the watercourse is navigable, then riparians are required to allow the general public to engage in certain surface uses (for example, floating) (see, for example, Mbaku 2009).

Prior Appropriation

The doctrine of prior appropriation developed in the United States during the mining boom that swept the western part of the country in the mid-1800s. Miners arriving in California in search of gold, for example, found themselves unable to assert riparian rights to water because they did not own any land. The miners resorted to a rule that they had applied in earlier disputes regarding access to minerals found on public lands—the "first in time, first in right" doctrine (see, for example, Gopalakrishnan, Tortajada, and Biswas 2005). Under this principle, the earliest or first miner to put the water to productive use automatically and implicitly had a right to continue using the water and to exclude use by others.

Taking into consideration the miners' customs and traditions regarding water use, early courts in the American West recognized these water rights. Eventually, the system of first in time, first in right became entrenched in the laws of most states of the American West and came to be interpreted this way: water rights belong "to anyone who puts water to a 'beneficial use' anywhere (on riparian or non-riparian land), with superiority over anyone who later begins using water" (Getches 1990, p. 6). Landownership, which is critical to the riparian system of rights, is not relevant to the first in time, first in right system, which has come to be referred to as prior appropriation.

Under the prior-appropriation system, a person need only put water to beneficial use and then comply with all necessary statutory requirements. Once that is done, the water user's right is perfected and remains valid for as long as he or she continues to exercise it.

Over the years, many countries with riparian systems have recognized rigidities inherent in this approach to water use—for example, commercial and agricultural activities, such as mining and farming, conducted on nonriparian lands could not always have access to the water. A much bigger problem was that development in many of these countries was often held hostage by riparian landowners who monopolized scarce water resources but were not willing to allow their rights to be interfered with for the pursuit of national development. However, if the prior-appropriation system is in force, water rights can be transferred "if it is shown that the ability of others to exercise vested rights is not impaired" (Getches 1990, p. 7).

In those American states that practice the prior-appropriation system, permits must be obtained from a designated administrative agency in order to appropriate water. These permits are issued taking into consideration regulations designed to protect the rights of other water users, as well as various public interest concerns.

In many African countries, the presence of Europeans during the colonial period has had a significant impact on the types of institutions that regulate water use in these countries. For example, in South Africa, as was the case in other parts of the continent, water use in the precolonial period was governed by customary law. Water law in South Africa during colonialism was influenced first by Dutch law and then by English law as the country was colonized by the Dutch and then by the English.[17] Along the way, the Anglo-American doctrine of riparian rights came to have a significant impact on water and water-use rights in South Africa and other countries colonized by Great Britain. In the postindependence period, some African countries have adopted new statutes to regulate water use.

17. D. D. Tewari, "A Detailed Analysis of Evolution of Water Rights in South Africa: An Account of Three and a Half Centuries from 1652 A.D. to Present," 2005, South African Water Research Commission (www.ajol.info/index.php/wsa/article/download/76851/67320); Hall and Fagan (1933).

7 The Nile Basin Initiative

In 1970 Egyptian president Anwar el-Sadat threatened to wage war with the Nile River's upstream riparian states, if necessary, to prevent any interference with the flow of water into the Nile River. "Tampering with the rights of a nation to water," he declared, "is tampering with its life; and a decision to go to war on this score is indisputable in the international community" (quoted in Kalpakian 2004, p. 55).[1] Egypt, of course, has gone to war with Ethiopia in the past, in an attempt to stop the latter from building structures on the Blue Nile that would negatively affect the flow of water from the Blue Nile into the Nile River (Kendie 1999).[2] It is important, then, that the eleven riparian states make a concerted effort to develop and adopt a legal and institutional framework that would enhance peaceful coexistence of the Nile River basin's riparian states and provide for equitable, reasonable, and sustainable utilization of the Nile's waters and other resources.

Although the upper riparian states have continued to argue that they do not recognize the Nile Waters agreements, they have indicated that they are willing and eager to cooperate with respect to efforts to resolve any conflicts associated with each riparian state's right to use the waters of the Nile River (Brunnée and Toope 2002, pp. 106–07). Given that the Nile River is an international watercourse, its management requires full and effective cooperation of all its riparian states.

1. Successive Egyptian leaders have made similar threats to those made by President Sadat. See, for example, Boutros-Ghali (1997).

2. For example, the Ethiopian-Egyptian War, 1874–76, which the Ethiopians won. See, generally, Degefu (2003) and Jonas (2011).

In 1999 the Nile River riparian states launched the Nile Basin Initiative (NBI) as a transitional scheme to provide them with a "comprehensive plan for transboundary cooperation over the Nile River" (Suvarna 2006, p. 450). The NBI employs a transitional arrangement to engage all stakeholders in negotiations to produce a permanent legal and institutional framework for Nile River governance. Such a framework, it was hoped, would recognize the importance of examining water resource and environmental management issues from an economic perspective.

Nile River Conflict

The Nile River is the lifeblood not just of Egypt but of a region that includes as many as ten other countries and a population of nearly 450 million people (World Bank 2011).[3] The Nile River stretches from Lake Victoria in the south to the Mediterranean Sea in the north. It flows through an area characterized by political instability, most of which can be traced to colonial interference in governance. One consequence of this instability is that the riparian states that emerged from the end of colonial rule have not been able to cooperate fully and effectively on issues critical to their economic, social, and political development, especially on their rights to the waters of the Nile River. As the region's population has continued to increase, there has been a significant expansion in the demand for water. Given that the Nile River is the most important source of water in the region, the increase in demand for water has meant that tremendous pressure is placed on the river to meet the region's domestic, commercial, and industrial water needs (Swain 2002, p. 298; Wiebe 2001, p. 734).

In the postcolonial period, industrial growth, increased demand for food and other primary commodities (for example, cotton), and significant increases in population have pushed the demand for the Nile River's waters past its available supply (Suvarna 2006, p. 453). Unfortunately, the Nile River basin does not currently have an effective legal and institutional framework to manage such a shortage effectively and provide for efficient utilization of the available water. Arun P. Elhance (1999, p. 59) has made the following determination:

> The minimum per capita water needs for an efficient, moderately industrialized nation have been estimated as 1,000 cubic meters per year, [and]

3. On July 9, 2011, after citizens of the various Sudanese provinces that were collectively referred to as Southern Sudan voted in a UN-supervised referendum to secede from the rest of Sudan and found an independent and sovereign country, South Sudan came into existence, bringing the number of Nile riparians to eleven.

> the annual per capita water availability between the years 1990 and 2025 is expected to drop from 1,070 to 620 cubic meters in Egypt, from 2,360 to 980 cubic meters in Ethiopia, from 590 to 190 cubic meters in Kenya, from 880 to 350 cubic meters in Rwanda, and from 2,780 to 900 cubic meters in Tanzania, based solely on the anticipated increases in the respective populations and on the current rates of water extraction.

This situation does not augur well for water allocation issues in the Nile River basin.

Presently, each riparian state in the Nile River basin has been left to manage what is essentially a regional problem from a national perspective.[4] Over the years, Egypt, for example, has sought to maximize its access to the waters of the Nile, paying very little attention to the needs and rights of the other riparian states (Waterbury 1979, p. 12). Egypt's decision to build the Aswan High Dam was based on its desire to improve its access to the waters of the Nile River; though the dam has, indeed, provided Egypt with many benefits, it has brought only misery to other riparians. The dam "has kept the river from flooding and depositing renewing sediment. . . . The delta has instead been inundated with catastrophic superlatives: It is among the world's most densely cultivated lands, with one of the world's highest uses of fertilizers and highest levels of soil salinity" (Theroux 1997, p. 8). The Aswan High Dam's construction, however, did not meet the demands of the upstream riparians for equitable and fair access to the waters of the Nile River.

The Nile Basin Initiative

The Nile Basin Initiative was signed in 1999 by all riparian states except Eritrea, and it is generally considered the "first basin-wide agreement regarding the Nile River" (Suvarna 2006, p. 455). Eritrea opted to participate in the NBI only as an observer. The main objective of the initiative was "to achieve sustainable socio-economic development through equitable utilization of, and benefit from, the common Nile Basin water resources."[5]

4. The Nile River is a shared resource, a regional public good whose management requires joint and coordinated efforts among its beneficiaries and those likely to be adversely affected by its exploitation.

5. Nile Basin Initiative, "The Shared Vision Objective" (www.nilebasin.org/index.php/about-us/nile-basin-initiative).

To direct the work of the initiative, the Council of Ministers of Water Affairs (of the Nile River basin states), which is the NBI's governing council, provided the following objectives for the strategic action program:[6]

—to develop the water resources of the basin in a sustainable and equitable way to ensure prosperity, security, and peace for all its peoples

—to ensure efficient water management and the optimal use of the resources

—to ensure cooperation and joint action between the riparian countries, seeking win-win gains

—to target poverty eradication and promote economic integration

—to ensure that the program results in a move from planning to action

As mentioned earlier, the emergence of the NBI signaled the embrace, by the Nile River basin riparian states, of a different and more democratic and accommodating attitude toward Nile River governance. The NBI thus has provided the riparian states with the opportunity to enter into a new and more sustained cycle in which a "broader integration of regional development in turn strengthens the relationships between the countries sharing international rivers, which further reinforces cooperation" (Patel 2003, p. vii; see also Suvarna 2006, pp. 45–56).

The NBI is not designed to serve as a permanent solution to the region's long-standing conflict over the allocation of water resources. Instead, it functions as a temporary or transient framework for "fostering dialogue, information exchange, and technology assistance as well as creating joint development initiatives in furtherance of transboundary goals" (Suvarna 2006, p. 456). If and when the Nile River basin riparian states do finally produce a viable framework for Nile River governance, it is expected that such a structure would provide the environment for improved economic and political cooperation, leading to significant improvements in economic growth and development, as well as peaceful coexistence, in the region.

A Shift in Relations among the Riparian States

The Nile Basin Initiative was expected to represent a significant change in both the nature and substance of cooperation between the Nile River basin states, as well as the substance of state-to-state relationships within the basin. The hope was that the NBI would replace the competitive nationalism that had characterized water allocation in the Nile River basin with a regionally based,

6. Nile Basin Initiative, "NBI Objectives" (www.nilebasin.org/index.php/about-us/nile-basin-initiative).

more inclusive and participatory approach not just to water allocation and use but also to the resolution of other problems associated with the Nile River (for example, construction of dams and other water diversion schemes, ecosystem restoration, and so on). Under the NBI, the riparian states were expected to participate fully and effectively in the negotiations to build a cooperative framework that would enhance development of the Nile River region.

The NBI is not the first such effort. By the 1980s, several Nile River riparian states had already indicated their interest in moving toward a more cooperative approach to resolving the region's conflicts, especially over the allocation of the waters of the Nile River. At the time, there emerged "an unofficial African regional grouping intended to serve as a platform for informal discussions regarding the overall economic development of the Nile Basin region" (Peichert 2000, p. 121). This unofficial consultation produced two important cooperative initiatives: Undugu and the Technical Cooperation Committee for the Promotion of the Development and Environmental Protection of the Nile Basin (TECCONILE).

Undugu—derived from the Swahili word *ndugu*, which means "brotherhood"—grew out of a proposal by Egypt to put together an informal organization that would use annual ministerial meetings to engage in discussions of various issues (water, resource management, economic development, technical and scientific cooperation among riparian states, and the like) associated with the utilization of the waters of the Nile River (Mekonnen 2010, p. 426). Undugu was embraced fully by Sudan, Uganda, Democratic Republic of Congo, and Central African Republic, a nonriparian state. Burundi, Rwanda, and Tanzania accepted the initiative later; Ethiopia and Kenya agreed to participate only as observers (Mekonnen 2010, p. 426).

Although Undugu performed well in providing a forum for the Nile River riparian states to share information and technology on various topics, it did not succeed in its effort to build cooperation among the riparians, in large measure because Egypt was not fully committed to the project, which it viewed as "an exercise in hegemonic influence" (Mekonnen 2010, p. 426). In fact, Egypt continued to carry out Nile River–related development and industrial projects without any consultation with the other riparian states, effectively defeating the very purpose for which Undugu was founded—improved cooperation among the riparian states, especially on issues related to the management of the waters and other resources of the Nile River (Mekonnen 2010, p. 426).

Undermined by Egyptian attitudes, Undugu eventually lost its relevance and was soon replaced by TECCONILE. The committee was established in

1992 by Egypt, Rwanda, Sudan, Tanzania, Uganda, and Zaire (now Democratic Republic of Congo). Other riparian states participated as observers (for example, Ethiopia), primarily because of the technical nature of the program. Those who established TECCONILE gave it a three-year lifespan and hoped that by the end of that period, the members would have created a permanent basin-wide institution (Mekonnen 2010, p. 426).

The committee proved to be an important launch pad for the next phase in cooperation in the Nile River basin: the Nile River Action Plan. The plan, developed and approved by the Council of Ministers of Water Affairs in February 1995, was instrumental in the founding of the NBI. Part of the plan, designed to foster regional cooperation, "envisaged the establishment of a basin-wide framework for legal and institutional arrangements which could not be implemented due to resource constraints and continued competitive behavior among the riparians" (Mekonnen 2010, p. 427). Ethiopia insisted that the principle of equitable entitlement of riparians to Nile River waters be included in the agenda as an issue to be given priority treatment. The principle was inserted, and the Nile Basin Cooperative Framework was duly "incorporated into the action plan, becoming, thus, a true progenitor of the NBCFA [Nile Basin Cooperative Framework Agreement] negotiated over the past decade under the aegis of the NBI" (Mekonnen 2010, p. 427).

Nile Basin Cooperative Framework Agreement

The Nile Basin Cooperative Framework Agreement represented the first attempt by the Nile River riparians to formally introduce the concept of equitable water allocation into discussions about Nile River governance.[7] The states that sponsored the agreement understood that if the Nile River were to become an instrument of economic growth and development, as well as industrial transformation in the region, there would have to be a permanent legal and institutional framework, established and agreed to by all the Nile River's riparian states through a democratic (that is, participatory and inclusive) process. The upstream riparian states, as well as most of the Nile River basin countries, recognized that the most important obstacle to Nile River basin cooperation was the bilateral treaties—the Nile Waters agreements—which they considered

7. Both Egypt and Sudan most likely agreed to participate in the Nile Basin Cooperative Framework Agreement negotiations to safeguard their rights and make certain that a final agreement would not be one that threatened what they saw as their historically acquired rights to the waters of the Nile River. Both countries have refused to sign the final agreement unless article 14(b), which limits such rights, is changed.

not binding on them. They argued strongly that a new, regionally based and inclusive agreement was needed to replace highly disputed bilateral treaties (Mekonnen 2010, p. 428; see also Salman 2012, pp. 17–29).

At the meeting of the Council of Ministers of Water Affairs of the Nile basin states in Entebbe, Uganda, in June 2007, the NBI presented the ministers with a draft of the Cooperative Framework Agreement (CFA) for discussion and approval. The attendees, however, were unable to make significant progress because of a key disagreement over article 14(b) of the agreement:

> Having due regard for the provision of Articles 4 and 5, Nile Basin states recognize the vital importance of water security to each of them. The States also recognize that cooperative management and the development of the waters of the Nile River System will facilitate achievement of water security and other benefits. Nile Basin states therefore agree, in a spirit of cooperation:
>
> (a) to work together to ensure that all States achieve and sustain water security
>
> (b) not to significantly affect the water security of any other Nile Basin State.[8]

Both Egypt and Sudan (the downstream riparian states) wanted article 14(b) amended to obligate all the riparian states "not to adversely affect the water security and current uses and rights of any other Nile Basin State."[9] The Council of Ministers of Water Affairs was unable to resolve the issue over the amendment and hence adopted the text of article 14(b) agreed on by all the riparians but also noted Egypt's proposed amendment to article 14(b) in an annex to the agreement. The group then referred the matter of water security to the heads of state and governments of the riparian states for resolution (Mekonnen 2010, p. 428).

Water Security

At its February 2002 meeting in Cairo, the Council of Ministers of Water Affairs established a negotiating committee to finalize a draft of the CFA. The negotiating committee, in performing its assignment, belatedly introduced the concept of water security into the agreement, supposedly to deal with the

8. Agreement on the Nile River Basin Cooperative Framework (www.internationalwaterlaw.org/documents/regionaldocs/Nile_River_Basin_Cooperative_Framework_2010.pdf); emphasis added.

9. Ibid., at Annex on Article 14(b). Also see Mekonnen (2010, p. 428).

thorny issue of existing treaties (that is, the Nile Waters agreements). Proponents of the water security provision claimed that its insertion into the CFA would provide an avenue for compromise, since it would make the existing treaties subordinate to international law (Mekonnen 2010, p. 430). Unfortunately, the insertion of the concept of water security introduced a seriously complicating factor into what had been, since the introduction of the CFA, a smooth transition to a cooperative environment for resolving Nile River governance issues. Although all parties were in agreement that some language about water security was appropriate, the upstream riparians did not agree to the effort by Egypt and Sudan to transform this language into the absolute protection of rights previously granted the two downstream riparians by the Nile Waters agreements. As noted, Egypt and Sudan wanted article 14(b) changed, to protect the rights granted them by the Nile Waters agreements.

Two justifications have been given for the decision to introduce the nonlegal concept of water security into the CFA: dealing with the issue of existing treaties and providing a conduit for so-called constructive ambiguity; the latter was expected to help bring all the "divergent riparian positions into compromise" (Mekonnen 2010, p. 430). The concept of water security, instead of helping relegate "existing treaties to the background in favor of the more dynamic and progressive principles of international water law" (Mekonnen 2010, p. 430), has actually prevented the NBI from reaching agreement on a legal framework that is acceptable to all riparians and hence defines rights and obligations as between all of the Nile River's riparians: both Egypt and Sudan have used the concept to secure for themselves absolute protection of their prior rights.

For millennia, Egypt has extracted nourishment from the Nile River. The country's near total dependence on the Nile River for economic and social survival, coupled with the river's erratic flow—a situation that has been made worse by climate change—has placed fear in the minds of Egyptians and their government that a possible diversion of the river upstream might sentence them to tremendous suffering and hardship. As a consequence, many Egyptian leaders have, in recent years, considered access to the waters of the Nile River to be an issue of national security (see, for example, Kendie 1999, p. 141).[10] In response to this fear, Egypt's leaders have pursued a public policy that has sought to monopolize access to the waters of the Nile River, a policy

10. Consider, for example, former Egyptian president Anwar Sadat's statement that "any action that would endanger the waters of the Blue Nile will be faced with a firm reaction on the part of Egypt, even if that action should lead to war." Quoted in Kendie (1999, p. 141).

that includes threats of military action, especially to upstream riparians, to ensure uninterrupted flow of water into the Nile River. Egypt, over the years, has come to believe that its hegemonic control over the Nile River is necessary not only for maintaining an uninterrupted flow of water into the Nile but also for economic, social, and political stability. However, cooperation among all the riparian states to develop an inclusive legal framework for Nile River governance would probably be a more effective way to resolve conflict over the Nile.

The binding effect of the Nile Waters agreements has been challenged. The theory of universal succession, discussed in chapter 5, has been used to support the argument that the 1929 Anglo-Egyptian treaty binds Kenya, Uganda, Tanzania, and Sudan (Mekonnen 2010, p. 432).[11] However, universal succession was abrogated by the Vienna Convention on Succession of States in Respect of Treaties: in the case of normal state succession, the Vienna convention applies the clean-slate principle (see chap. 5 in this volume).

Nevertheless, the "dispositive," "real," or "localized" treaties exception has been invoked to support a claim for the continuing binding force of the 1929 Anglo-Egyptian treaty on Britain's former colonies in East Africa. According to this exception, treaties such as the 1929 agreement survive the impact of succession and bind the successor state (Mekonnen 2010, p. 433). The normative validity of the exception, however, has been challenged (see, for example, Zeleke 2005, pp. 124–34), and the types of treaties to which the exception applies have been the subject of dispute (Mekonnen 2010, p. 433). From the time the agreement was concluded, Egypt had indicated that it was a temporary arrangement that would allow for the governance of the Nile River until critical issues such as the political future of Sudan had been resolved, after which policymakers could undertake the process of determining the future of the Nile River. Hence the Anglo-Egyptian agreement was designed to be "neither a real nor a dispositive agreement" (Okidi 1980, p. 423).

Although treaties dealing with boundary matters may fall under the exception, "the claim for the transmission of colonial era fluvial treaties onto the independent successor states has no legal foundation" (Mekonnen 2010, p. 433). The clean-slate principle, as defined by the Vienna Convention on Succession of States in Respect of Treaties, stipulates that a newly independent state "is not bound to maintain in force, or to become a party to, any treaty by reason only of the fact that at the date of the succession of states the treaty

11. Exchange of Notes between His Majesty's Government in the United Kingdom and the Egyptian Government in Regard to the Use of the Waters of the Nile River for Irrigation Purposes (with Seven Diagrams), Cairo, May 7, 1929, L.N.T.S. 2103.

was in force in respect of the territory to which the succession of state relates."[12] Of course, there have been a number of cases in which newly independent countries have voluntarily, and for self-interest, exercised their right to opt in and accept to take over the rights and obligations of their predecessors by agreeing to be bound by the treaties those predecessors had concluded (Mekonnen 2010, p. 434).

This opt-in or "operational" theory is in line with positions that had been followed by Kenya, Tanzania, and Uganda and refined, formulated, and proclaimed by Julius Nyerere in 1961 when he was serving as premier of an autonomous but not yet independent Tanganyika. The Nyerere doctrine, as it is called, categorically rejects "any categorization of international obligations which a successor state might have to accept or reject only because of the nature or type of the obligation," taking into consideration the demands and requirements of international law (Makonnen 1986, p. 123; see also Makonnen 1983). The primary impetus to the Nyerere doctrine is the assertion by many African countries that though they respect international law, they should not be forced to accept colonial treaties (Mekonnen 2010, p. 434).

Other former British colonies in East Africa, notably Uganda, followed Tanzania's example and rejected the 1929 agreement as well, on essentially the same grounds (Mekonnen 2010, p. 434). Sudan, though a beneficiary of the 1929 agreement, also rejected the agreement, in a proclamation made in 1958, two years after it gained independence from Britain and Egypt (Brunnée and Toope 2002, p. 167; also see Maluwa 1986) and effectively forced Egypt to renegotiate the 1929 agreement.

On November 8, 1959, Egypt and Sudan signed a bilateral agreement in Cairo, which called for the two countries to have full control and exclusive utilization of the waters of the Nile River.[13] This treaty is examined in chapter 4 of this volume. Although this agreement was lauded as the first cooperative effort between two independent riparians in the Nile River basin, it allowed Egypt and Sudan to have a complete monopoly over the waters of the Nile River. Nevertheless, such monopoly control was not anchored on sound legal ground because the agreement on which it was based "is a typical bilateral agreement subject to the pacta tertiis nec nocent nec prosunt rule of treaty law which, therefore, has no binding force on the other riparian States"

12. Vienna Convention on Succession of States in Respect of Treaties, Vienna, August 23, 1978, U.N.T.S. 1946, art. 16.

13. This treaty is officially referred to as United Arab Republic and Sudan Agreement (with Annexes) for the Full Utilization of the Nile Waters, Cairo, November 8, 1959, 6519 U.N.T.S. 63.

(Mekonnen 2010, p. 435). According to this rule, treaties neither impose obligations nor confer rights on third-party states. Given that neither the 1929 Anglo-Egyptian treaty nor the 1959 bilateral agreement between Egypt and Sudan have any binding effect on the upstream riparians, those states are not required to respect any rights created by these treaties.

An agreement on an allocation scheme among the Nile River basin states should emphasize equity and fairness. Egypt, of course, will suffer a reduced legal share of the Nile River waters and would most likely protest bitterly. However, equity and fairness, critical for maintaining peaceful coexistence in the Basin, require that allocations to Egypt be reduced to make allowance for actualization of the rights of other riparians, especially given the impacts of climate change and population increases in the region, including Egypt.

8 The Cooperative Framework Agreement: A New Legal Regime for the Nile River

Can the Nile Basin Initiative's Cooperative Framework Agreement (CFA) serve as the new legal instrument for the governance of the Nile River basin? Does the agreement satisfy the requirements of an effective international watercourse law? For the CFA to function as an appropriate legal instrument for governing the use of the waters of the Nile River, it must be acceptable to all of the watercourse's riparians, properly define each riparian's rights and obligations, and provide them with appropriate tools to peacefully resolve conflict, especially that related to the distribution of the Nile waters.

Egypt and Sudan, both downstream riparians, are essentially consumers and not suppliers of waters to the Nile River. The populations of both countries are dependent, to a significant extent, on the waters of the Nile. Egypt's population, which stood at 84 million in 2012 (World Bank 2012, p. 392), resides almost exclusively in three areas: the cities of Cairo and Alexandria and the densely populated communities along the Nile River. Water from the Nile provides for as much as 96 percent of Egypt's renewable water needs (Hefny and el-Din Amer 2005, p. 42); 85 percent of the population of Sudan (now two countries, the Republic of Sudan and South Sudan) depends on the Nile River (el-Tom Hamad and el-Battahani 2005, p. 28; see also Ibrahim 2011, p. 287).

Another Look at the Nile River Basin Cooperative Framework

Beginning in 1999, the Nile River basin states, working under the auspices of the Nile Basin Initiative, began negotiations to design a legal framework that would enable fair and equitable allocation of the waters of the Nile and, they hoped, might become the international watercourse's legal regime. Negotiations for the development of that legal framework, which came to be called the Cooperative Framework Agreement, were carried out under the leadership of the Council of Ministers of Water Affairs of the Nile River basin countries. As is discussed in chapter 7, there was a disagreement between the upstream and downstream riparian states over article 14(b). Approval of the alternative wording proposed by Egypt and Sudan would have served to entrench the concept of absolute protection of prior rights, as claimed by Egypt and the Republic of Sudan. All the upstream riparian states rejected the proposal.

The Nile River Basin Commission was supposed to resolve the wording of article 14(b) within six months after its establishment. Unfortunately, all efforts to resolve this issue have been fruitless. Beginning on May 10, 2010, the instrument was open for signature by member countries. The closing date was fixed for May 14, 2011.

On May 10, 2010, five of the upstream riparian states—Ethiopia, Kenya, Rwanda, Tanzania, and Uganda—signed the Cooperative Framework Agreement. Egypt and Sudan, the main beneficiaries of the Nile Waters agreements, strongly objected to the action taken by the five upstream riparians. On February 28, 2011, Burundi, another upstream riparian, also signed the CFA. In 2013 the Ethiopian parliament ratified the treaty.

Both Sudan and Egypt are located primarily in the Sahara desert, and hence they cannot survive, let alone achieve any significant level of development, without the Nile River. It is no wonder that Egyptian leaders, past and present, have come to recognize access to the waters of the Nile River, as well as the viability of the Nile River itself, as an issue of national security (Mekonnen 2010; Degefu 2003). Although the Democratic Republic of Congo has indicated its interest to sign the agreement, it has not yet done so.

Since the 1920s, while Egypt and Sudan have harvested and used the Nile River to turn vast stretches of desert into productive farmland and to provide their citizens with hydroelectric power, the upstream riparian states have made little use of these waters (see, for example, Ibrahim 2011, pp. 287–88; Amdetsion 2008, pp. 9–10). The failure of the upstream riparians to make effective use of the waters of the Nile can be attributed to many factors, some of them external (for example, colonial policies) and others internal (for example, lack of capacity to effectively build and operate necessary irrigation schemes). Whereas Egypt, over the years, has been able to effectively utilize the 55.5 billion cubic meters of Nile water a year allocated to it by the 1959 bilateral agreement,[1] the upstream riparians have used virtually none. For example, over the years, Ethiopia has been able to irrigate only 190,000 hectares of land out of 3,637,000 hectares of irrigable land available to the country (Amdetsion 2008, p. 9).

Other upstream states have fared no better. Hence the CFA is considered an important first step in securing water rights for all riparian states. Within such an institutional environment, the upstream riparians can undertake the necessary investments to harness Nile River waters for national development such as the Grand Ethiopian Renaissance Dam. Egyptian and Sudanese authorities, however, have continued to argue that they will not sign the CFA, stating that they would not become signatories to any agreement to govern the Nile River "unless they are first guaranteed an exact share of the water."[2]

Cooperative Framework Agreement as the International Watercourse's Legal Regime

The main objective of the upstream riparian states in promoting the enactment of the Cooperative Framework Agreement was to produce an effective legal instrument for Nile River governance. They hoped that with the institution of this new legal framework, both Egypt and the Republic of Sudan would abandon the bilateral treaties, which they claim have granted them historic rights to the waters of the Nile River.

As it stands, all members of the Nile Basin Initiative have agreed on the provisions of the CFA with the exception of two critical issues: whether the new treaty, when it eventually becomes operational, will nullify all previous Nile River–related agreements, whether bilateral or otherwise, and produce an allocation and utilization mechanism for the waters of the Nile River that is acceptable to all riparian states, and whether the new agreement adequately addresses the issue of water security in a way that is acceptable to both upstream and downstream riparians. During the negotiations to structure

1. United Arab Republic and Sudan Agreement (with Annexes) for the Full Utilization of the Nile Waters, Cairo, November 8, 1959, 6519 U.N.T.S. 63.

2. BBC News, "East Africa Seeks More Nile Water from Egypt," *BBC News*, May 14, 2010 (http://news.bbc.co.uk/2/hi/africa/8682387.stm). See also N. Jemal, "Sudan: South Sudan Set to Sign New Nile Agreement," Al Jazeera, June 20, 2013 (www.aljazeera.com/news/africa/2013/06/201362075235645727.html).

the CFA, the upstream riparian states wanted the new legal regime to supersede all previous agreements, including the Nile Waters agreements, and provide all riparians with a legal mechanism capable of achieving equity and fairness in allocation, as well as meeting the needs of economies searching for ways to improve the living standards of their citizens. The downstream riparians (Egypt and Sudan), however, insisted that the CFA specifically and explicitly recognize and accept as legally legitimate and binding all previous agreements related to the allocation and use of the waters of the Nile River. None of the CFA's more than forty articles tackles the important issue of previous agreements: hence the fate of the 1929 Anglo-Egyptian treaty and the 1959 bilateral agreement between Egypt and Sudan is not specifically addressed by the CFA. The new agreement, however, focuses on a new controversy that has arisen within the Nile River Basin—that is, water security.

The drafting committee's official position was to leave the issue of previous agreements for deliberation at a later time (Ibrahim 2011, pp. 302–03). However, Ibrahim (2011, p. 303) argues that "although the official position held by the drafting committee is to leave the issue of the colonial treaties for later deliberation, it is very likely the new treaty will have the legal effect of annulling the previous treaties" (also see article 59 of the Vienna Convention on the Law of Treaties).[3] However, the problem that the Nile Basin Initiative faces in achieving such an outcome is that Egypt and Sudan have indicated in no uncertain terms that they will not sign any agreement that is designed to either annul the Nile Waters agreements or abrogate the rights granted them by these bilateral treaties.

One could ask, then, why the upstream riparian states would proceed with the development of an agreement that is not likely to have the desired legal effect because Egypt and Sudan have refused to be signatories to it. Is the CFA basically an instrument designed solely to improve the negotiating environment in the Nile River basin, forcing Egypt and Sudan to give up the rights granted them by the Nile Waters agreements and engage the upstream riparian states in robust negotiations to craft a new legal framework that is acceptable to all of the Nile River's riparians?

Article 30 of the CFA states that "upon the entry into force of [the CFA] the Commission shall succeed to all rights, obligations and assets of the Nile Basin Initiative." Article 42 of the CFA states that "the present Framework shall enter into force on the sixtieth day following the date of the deposit of the sixth instrument of ratification or accession with the African Union."

3. Vienna Convention on Succession of States in Respect of Treaties, Vienna, August 23, 1978, U.N.T.S. 1946.

Given the fact that six countries, all upstream riparian states, have already signed the accord, it is possible that if the legislatures of the six countries ratify the CFA and deposit it with the African Union, the agreement will enter into force within a few months.[4]

But what will be the legal status of such an agreement if both Egypt and Sudan refuse to sign it? If the most important, and perhaps the only, reason the upstream riparian states wanted to develop the CFA was to resolve the long-standing conflict between the upstream and downstream riparian states over rights to the Nile River's waters, then the CFA as it stands fails to accomplish that goal. Thus even if the agreement does eventually enter into force as mandated by article 42, it would lack the legal ability to resolve the conflict that has consumed issues of water allocation in the Nile River basin since the 1920s. Without the participation of Egypt and Sudan, the CFA would simply be unilateral action on the part of the upstream riparian states and would certainly not bind the downstream riparian states.

Some commentators have argued that the importance of the CFA lies in the possibility that it could be used by the upstream riparian states to "diplomatically corner the lower states" and encourage them to abandon their insistence that the Nile Waters agreements be retained (Ibrahim 2011, p. 305). Although it is not clear whether upstream riparians such as Ethiopia expect Egypt to eventually abandon its hegemonic control of the Nile River and opt for a legal regime that guarantees fairness and equity, Ethiopian leaders have intimated that the main problem is Egypt and not Sudan (Ibrahim 2011, p. 305). There are expectations that both the Republic of Sudan and South Sudan will eventually sign the Cooperative Framework Agreement.

South Sudan authorities have expressed their opposition to the 1959 bilateral agreement between Egypt and Sudan and the country's intention to sign the Cooperative Framework Agreement. As the country's water and irrigation minister, Paul Mayom Akech, has stated, "South Sudan does not recognize—and underline does not recognize—the content of the 1959 agreement." The minister reiterated that "having been under Sudan at the time [that is, in 1959] we could not say anything, today we say, we have nothing to do with this agreement."[5]

Meanwhile, Eritrea, which historically has sided with Egypt in disputes between Egypt and Ethiopia, has announced its support for Egypt's position

4. The Ethiopian Parliament ratified the Cooperative Framework Agreement in June 2013 and encouraged the other countries to ratify it as well.

5. Quoted in Machel Amos, "South Sudan Rejects the Colonial Nile Waters Agreement," Meles Zenawi Memorial (www.meleszenawi.com/south-sudan-rejects-the-colonial-nile-waters-agreement).

on the issue of the type of legal regime that should be adopted to govern the Nile River.[6] Eritrean authorities support Egypt's desire to retain the water allocations assigned to it by the Nile Waters agreements.

Some scholars have argued that, over the years, Egyptian officials have employed their diplomatic skills to pressure the international financial and donor community not to grant financing for projects on the Nile River or its tributaries that might affect Egypt's effective utilization of the waters allotted to it by the Nile Waters agreements. In fact, according to John Waterbury (2002, pp. 71–72), Egypt has pursued a foreign policy that has deliberately sought to discourage international and multilateral organizations such as the World Bank and the International Monetary Fund from acting favorably toward any of the upstream riparian states in matters related to the Nile River.[7] Abadir M. Ibrahim (2011, p. 306) argues that "because of Egypt's strategic and economic importance, it is also unlikely that upper riparian states will find . . . donors or lenders who will make exception to Egypt's influence." As a consequence, upstream riparian states have found it increasingly difficult to access the financial resources that they need to develop projects on the Nile River.[8] Without a single comprehensive legal instrument acceptable to all riparian states, it will be difficult for the Nile River basin to adopt "an integrated development approach" or undertake "basin-wide or holistic development and control" (Jacobs 1993, p. 120).

The problem with the position held by these scholars with regard to Egyptian foreign policy is that the World Bank and other multilateral financial institutions may have arrived at their positions regarding the financing of Nile River–related projects independent of any influence from Egypt or its benefactors. In fact, as a financial institution, the World Bank most likely considered both economic and political risk factors, which would have included the lack of an accepted legal framework, as well as the disputed water rights, in any such financing and arrived at the conclusion that these investments were simply too risky.

The question in the Nile River basin today is not whether to change the status quo but how to do so. One can look at the Cooperative Framework

6. Tesfa-Alem Tekle, "Eritrea Supports Egypt's Position over Nile Water Dispute," *Sudan Tribune*, April 18, 2013 (www.sudantribune.com/spip.php?article46276).

7. See, for example, Seifulaziz Milas, "Ethiopia: Nile Waters Diplomacy and the Renaissance Dam," *African Arguments*, October 3, 2012 (africanarguments.org/2012/10/03/ethiopia-nile-waters-diplomacy-and-the-renaissance-dam-by-seifulaziz-milas/).

8. In fact, Ethiopia has been forced to rely on domestic sources to finance its Grand Ethiopian Renaissance Dam.

Agreement as a tool to achieve two important and related objectives: encourage both Egypt and Sudan to consider giving up the water rights granted them by the Nile Waters agreements and provide the mechanism for the development of a new legal regime that is acceptable to all riparians and provides for the equitable and reasonable allocation of the waters of the Nile River. With a stable legal environment, the basin can finally achieve the economic and political grounding sufficient to enhance the riparians' ability to source the necessary financing for the construction of infrastructure for the effective use of the Nile River for national development (Ibrahim 2011, p. 310).[9]

9. As argued by Michael Hammond (2013, p. 2), the World Bank has stated that "increased investment in multipurpose water infrastructure would make Ethiopia more 'water-resilient,' and promote long-term economic growth." However, Hammond (2013, p. 2) notes, "The World Bank and other international donors have refused to support the [Grand Ethiopian Renaissance Dam] project, and the Ethiopian Government is attempting to finance the project through a national bond." Again, the decision by the World Bank to deny funding to Ethiopia may not necessarily be based on sympathy for Egypt, but on economic and financial considerations.

9 Egypt, Ethiopia, and the Nile River

For thousands of years, Ethiopian authorities have recognized the importance of the Nile River to the survival and vitality of Egyptian agriculture and hence human survival and development in Egypt. In fact, Ethiopian monarchs, "who had a fair awareness of the vitality of the Nile floods for the survival of Egypt, used it as a rough diplomatic whip to pressurize their Egyptian counterparts on matters which then constituted their primary concerns" (Mekonnen 2010, p. 423; see also Erlich 2002). Over the years, these threats have defined the relationship between Egypt and Ethiopia and the other upstream Nile River riparian states. Egyptians came to see Ethiopia as the most important threat to Egypt's survival as a nation. For quite some time now, Egyptians have expressed the fear that upstream riparian states, including especially Ethiopia, can significantly and negatively influence the livelihoods of the people who live downstream, particularly Egypt's population, which is located almost exclusively along the banks of the Nile River and in the Nile delta (Collins 2002, p. 22).

During the colonial period, British authorities in the Nile River basin pursued a policy that called for the effective consolidation of the control of the entire basin to make certain that there would be no interruption of the flow of water into the Nile River. Thus British colonialism created a "new reality that would have profound implications for inter-riparian relations long after [its] departure" (Yohannes 2008, p. 35). In fact, British colonial officers were so concerned about an adequate supply of water to their cotton interests in Egypt that London supported the Italian invasion of Ethiopia in the early 1890s in order to weaken the latter and enhance the efforts of the colonial government in Cairo to prevent authorities in Addis Ababa from interfering with the flow of water from the Blue Nile into the Nile River (Marcus 1963, pp. 81–89; see also Dellapenna 1994, pp. 27–56).

The decision by the British government under Lord Salisbury in 1889 to have control not just over Egypt but also over the Nile River meant that the colonial government in Cairo had to find ways to minimize the possibility that Ethiopian authorities might interfere with the flow of the Blue Nile into the Nile River. As stated by Lord Salisbury, "Since upon its mountains fell the abundant rains which furnish the very life of Egypt and the eastern Soudan, it is possible for the state in possession of these mountains to flood the Valley of the Nile or make of it a blistering desert at will" (quoted in Marcus 1963, p. 81). Harold G. Marcus (1963, p. 81) argues that British policy in the 1890s was "designed to close the Nile Valley to other European powers from the east as well as from the west. As a counterpoise to the French, who were her most serious rivals in the Nile Valley, and to the Emperor Menelek's stated policy of expansion, the British Government supported Italian colonial ambitions in Ethiopia."

Britain's efforts to ensure a steady and reliable flow of water into the Nile River were designed, of course, not to maximize Egyptian interests but to cater to British agricultural investments and other related interests in the Nile River basin. Independent Egypt essentially adopted colonial Britain's approach to the allocation of the Nile waters—hegemonic control and competition—and opposed or tried to minimize efforts by upstream riparian states to forge a cooperative framework for dealing with Nile River issues. This Egyptian approach to Nile River governance is evident in the 1959 bilateral agreement with Sudan, which not only granted Egypt most of the waters of the Nile River but also granted Cairo veto power over development projects on the Nile River and its tributaries.[1] Thus until recently, issues of water allocation in the Nile River basin came to be dominated by the political economy in Egypt and Sudan, the main users of the Nile River's waters, and Ethiopia, whose Lake Tana is the most important source of the Nile River. Understanding the tumultuous relations between Egypt, Sudan, and Ethiopia is critical

1. See the 1959 bilateral agreement between Egypt and Sudan (United Arab Republic and Sudan Agreement [with Annexes] for the Full Utilization of the Nile Waters, Cairo, November 8, 1959, 6519 U.N.T.S. 63). Of course, as mentioned earlier, being a bilateral agreement, the treaty could only grant rights as between those states that were actually bound by the treaty—that is, Egypt and Sudan—and did not do so relative to parties (that is, the upstream riparian states) that were not bound by the treaty.

to appreciating the nature of the disagreements that exist in the Nile River basin regarding the allocation of water.

However, in this chapter, we are primarily concerned with the relationship between Egypt and Ethiopia, the two most powerful states in the Nile River basin (Collins 2002, p. 214). Ethiopia is the source of virtually all of the water that flows into the Nile River. Egypt is currently the user of most of these waters and insists that it has a historical right to harvest for its use virtually all of the Nile River's waters (Collins 2002, pp. 214–15).

Over the years, the relationship between Egypt and Ethiopia has been framed, to a great extent, by Egypt's fear that Ethiopian authorities could engage in activities that might severely undermine the flow of water from the Blue Nile into the Nile River. Occasionally, Egyptian leaders have threatened force against Ethiopia to avoid any manmade interruptions to the flow of water from the Ethiopian highlands into the Nile River. In fact, historians state that Egypt invaded and conquered Sudan in 1820 largely in an effort to "secure control over the entire Nile system" (Kendie 1999, p. 145). The conquest of Sudan, however, was expected to serve "as a stepping-stone to the increased appearance of Egyptian soldiers in the western frontiers of Ethiopia, and to the subsequent Egyptian occupation of Kasala in 1834, Metema in 1838, Massawa in 1846, Kunama in 1869, and Harar in 1875" (Kendie 1999, p. 145; see also White 1899). Through the years, Egyptian rulers have pursued policies designed to subjugate Ethiopia, militarily or otherwise, to gain or maintain control over the waters of the Nile River—specifically, Egypt's objective has been to control the Ethiopian-based Blue Nile. Consider this statement by Werner Munzinger, who was in the service of Egyptian ruler Khedive Ismail from 1863 to 1879: "Ethiopia with a disciplined administration and army, and a friend of the European powers, is a danger for Egypt. Egypt must either take over Ethiopia and Islamize it, or retain it in anarchy and misery" (Rubenson 1976, p. 200; also quoted in Kendie 1999, p. 145).

Khedive Ismail's military expeditions into Ethiopia did not produce the results that he had expected or anticipated. Between 1875 and 1876, the Egyptian leader launched attacks against Ethiopia. During action that lasted from November 14 to 16, 1875, Ethiopian soldiers defeated 2,500 Egyptian fighters at the Battle of Gundet (Gabre-Selassie 1975, pp. 54–83; see also Dunn 2005). At the Battle of Gura, which lasted from March 7 to 9, 1876, Ethiopian forces routed 12,000 Egyptian soldiers, despite the fact that Egyptian authorities had enlisted the help of foreign mercenaries—notably William M. Dye, a former military officer who had fought with Union forces in the American Civil War and had joined the Egyptian army in 1876 (see, for example, Vivian 2012).

Despite these defeats at the hands of Ethiopian forces, Egyptian authorities continued to organize and carry out raids against Ethiopia until the arrival of British colonialists in the late nineteenth century (Kendie 1999, p. 146).

After the British took control of Egypt in 1882, following the defeat of Egyptian forces at Tel el Kebir, the new rulers recognized the importance of the Blue Nile to Egypt and began to make arrangements to engage in negotiations with Ethiopia's rulers to make certain that the latter did not interfere with the flow of water into the Nile River, thereby harming British agricultural and other interests in the region. In 1902 British authorities in London sent the diplomat Sir John Harrington to Addis Ababa to meet with Ethiopian authorities and begin negotiations on various issues, especially the allocation and utilization of the waters of the Nile River. Harrington's work in Ethiopia produced the Anglo-Ethiopian treaty of 1902 (Kendie 1999, p. 146). As its full name implies, the 1902 treaty's most important objective was to establish the boundary between Sudan, then a British colony, and Ethiopia, an independent and sovereign kingdom.[2] Nevertheless, article 3 of the treaty was devoted to issues associated with the allocation of the waters of the Nile River.[3]

Ethiopian leaders, of course, were quite aware that the country needed to use its own water resources for national development. As Daniel Kendie (1999, p. 146) argues, "This fact alone would provide sufficient grounds for some to invalidate the binding force of the agreement." There was no need, however, for Ethiopians to seek to invalidate the agreement because neither the British

2. Treaties between the United Kingdom and Ethiopia and between the United Kingdom, Italy, and Ethiopia, Relative to the Frontiers between the Soudan, Ethiopia, and Eritrea, Signed at Addis Ababa, May 15, 1902, Treaty Series, No. 16 (1902) (http://treaties.fco.gov.uk/docs/pdf/1902/TS0016.pdf).

3. Article 3 is reproduced here to emphasize the problem associated with the mistranslation of the word *arrest* into the Ethiopian language. Here is the text of the article: "His Majesty the Emperor Menelik, King of Kings of Ethiopia, engages himself towards the Government of His Britannic Majesty not to construct or allow to be constructed, any works across the Blue Nile, Lake Tana or the Sabot, which would *arrest* the flow of their waters into the Nile except in agreement with His Britannic Majesty's Government and the Government of the Sudan" (see Kendie 1999, p. 146, emphasis added). The emperor is said to have signed the agreement owing to a mistranslation of the word arrest, which had been translated into the Amharic version of the treaty as "stop," implying that the emperor could, under the terms of the treaty, harvest and use the waters of the Blue Nile for national development, as long as doing so did not "stop" the flow of water into Egypt. See, for example, J. H. S. Lie, "Supporting the Nile Basin Initiative: A Political Analysis 'Beyond the River,'" Norwegian Agency for Development Cooperation, 2010 (http://academia.edu/2243972/Supporting_the_Nile_Basin_Initiative_A_Political_Analysis_Beyond_the_River).

Parliament nor the Ethiopian Crown Council ever ratified the treaty and hence it never became law in either country (Kendie 1999, p. 146).

Continued British interests in the Nile River basin and its resources were also reflected in the exchange of correspondence between Britain and Italy. Britain engaged in a secret deal with Italy so that the former could build certain infrastructure on the Nile River and its tributaries, including a barrage at Lake Tana, to enhance British trade in the region. Ethiopian authorities condemned the secret agreement (Kendie 1999, p. 147). And the 1929 Anglo-Egyptian treaty secured necessary water for British cotton interests in the Nile River basin.[4]

With reference to the 1929 treaty between Britain and Egypt, the former could not have been representing Ethiopia since Ethiopia had never been a colony of Britain or of any other European country, and so, as Ethiopian leaders have argued, the country is not legally bound by this treaty.[5] Perhaps more important is the argument advanced by Ethiopian authorities that it was hardly fair, reasonable, or equitable for Egyptian authorities to demand that Ethiopia accept an agreement that granted Egypt authority to control development policy in Ethiopia—for, by subjecting irrigation and other projects on the Blue Nile to prior approval and oversight by Egypt, the treaty effectively made Egypt an overseer of development policies in Ethiopia. The 1929 Anglo-Egyptian treaty, Ethiopian authorities argued, was designed exclusively to protect Egyptian interests, and no effort was made to account for the development needs of the upstream riparian communities, including, of course, those in Ethiopia. Thus as early as the 1950s, Ethiopia had "asserted and reserved, then and in the future, the right to utilize the waters of the Blue Nile without recognizing any limitations on its freedom of action. It also invoked its new economic needs as grounds for its release from old treaty obligations" (Kendie 1999, p. 148; see also Abdalla 1971, pp. 329–41).

Ethiopia's protests against the Nile Waters agreements were quite specific and included diplomatic correspondences. The Ethiopian Aide Memoire of September 23, 1957, which was circulated in the various diplomatic missions in Cairo, is an example of such efforts to inform Egyptian authorities and the international community of the country's desire to distance itself from what it believed were agreements designed to abrogate its rights to the exploitation and use of its own natural resources. The Aide Memoire states specifically that "Ethiopia has the right and obligation to exploit its water resources, for

4. Exchange of Notes between His Majesty's Government in the United Kingdom and the Egyptian Government in Regard to the Use of the Waters of the River Nile for Irrigation Purposes (with Seven Diagrams), Cairo, May 7, 1929, L.N.T.S. 2103.

5. Italy did occupy Ethiopia from 1936 to 1941.

the benefit of present and future generations of its citizens [and] must, therefore, reassert and reserve now and for the future, the right to take all such measures in respect of its water resources" (quoted in Whiteman 1964, pp. 1011–12; also see Kendie 1999, p. 148).

Ethiopian Aspirations for the Blue Nile

Over the years, the Ethiopian government has expressed an interest in building a dam on Lake Tana, the source of the Blue Nile. Feasibility studies of building a dam at the source of the Blue Nile were conducted, including one that was carried out by the J. G. White Engineering Corporation of New York in the 1920s (Kendie 1999, p. 149; McCann 1981, pp. 667–96). More recently, studies sponsored by the government of Ethiopia and other parties have determined that the terrain of the Blue Nile is particularly suitable for the building of dams to generate electricity that could meet the power needs not just of Ethiopia but also of other communities in the Nile River basin and even the Arabian Peninsula (Kendie 1999, p. 149). Expert studies have revealed that "the amount of water available to the downstream riparian states would not be affected, even if Ethiopia were to implement the Blue Nile Plan, drawing off six km^3, Egypt and the Sudan would still benefit from the construction of the reservoirs within Ethiopia" (Kendie 1999, p. 150; see also Guariso and Whittington 1987, pp. 105–14).

Scholars have investigated why Ethiopia has not, until recently, put its plans into effect (see, for example, Kendie 1999; Collins 2002). First, agriculture in Ethiopia has historically relied primarily on rain-fed irrigation methods. Second, Egypt's continuous and persistent efforts to destabilize Ethiopia have forced Addis Ababa to devote a significant part of its scarce resources to national security and defense. Hence Ethiopia's use of the waters of the Blue Nile has been minimal. During the past several decades, however, Ethiopian officials, faced with a deteriorating food situation that includes the possibility that the country might not be able to feed its burgeoning population if it continues to rely on existing agricultural practices, have made concerted efforts to explore ways to use the waters of the Blue Nile for improved agricultural production. Part of that effort has involved the increase of government investment in irrigation infrastructure to allow farmers to more effectively benefit from the waters of the Blue Nile (Kendie 1999, p. 150; Howell and Allan 1994). Addis Ababa, of course, is also aware of the country's need for more reliable sources of clean, reliable, and affordable energy, particularly electricity, for economic growth and development.

Seeking Ways to Cooperate on the Blue Nile

The Blue Nile is an important watercourse to both Egypt and Ethiopia. Both countries would benefit from cooperation in the design of a legal framework that would allow for the fair and equitable utilization of the river's waters. Unfortunately, the two countries have not been able to engage in the type of cooperation that would enhance the creation, by all riparians, of an effective legal regime. Instead, officials in Cairo and Khartoum have continued to insist that the rights granted them by the Nile Waters agreements be recognized and respected by all Nile River riparians.

Some history might throw light on the troubled relationship between Egypt and Ethiopia regarding the Blue Nile. After Italy's eviction from Eritrea in 1941, Egyptian authorities hoped to claim the territory and use it as a base from which to destabilize Ethiopia and make sure that no activities were conducted on the Blue Nile (for example, the construction of dams) that would interfere with the flow of water into the Nile River (Kendie 1999, pp. 153–54). After its efforts to claim Eritrea had failed, Egyptian president Gamal Abdel Nasser engineered a campaign that he claimed was designed to unite all the countries of the Nile River basin. But it soon became clear that Nasser's concept of unity meant the bringing of all Nile River Basin communities under Egyptian control (Kendie 1999, p. 153).

Shortly thereafter, Nasser and his lieutenants embarked on a foreign policy that had the destabilization of Ethiopia and the undermining of its government as a top priority. Egyptian plans to destabilize Ethiopia would have been illegal under international law, so Egypt had to rely on more covert efforts (Kendie 1999, p. 154). An important part of Egypt's plan to destabilize the government in Addis Ababa involved providing military training for Eritreans who would later lead the irredentist province's war of independence. Eventually, broadcasts from Radio Cairo began to encourage Eritreans not just to undermine the government of Emperor Haile Selassie but also to wage war against Ethiopia for Eritrea's secession and eventual independence (Kendie 1999, p. 154).[6]

In 1962 the government of Ethiopia abrogated Eritrea's status as a UN-mandated federated state, dissolving the Federation of Ethiopia and Eritrea, which had been established by a United Nations resolution, on November 15, 1962, and converting Eritrea into a province of Ethiopia. Shortly afterward,

6. Even without the encouragement and support of Egypt, Eritreans would still have moved ahead with their struggle for self-determination and eventual independence from Ethiopia.

Egyptian authorities established offices in Cairo for Eritrean freedom fighters who were seeking secession and subsequent creation of an independent Eritrean state. Soon, Egypt would make Eritrea's struggle against Ethiopia an Arab League issue, befitting the organization's full support. It was claimed that Eritreans were "Arabs and overwhelmingly Muslim . . . struggling against the forces of 'Zionism,' 'American imperialism,' and 'Ethiopian colonialism'; that in violation of its status as a member of the Non-Aligned Movement, Ethiopia had provided the United States with military bases to spy on the Soviet Union and the Arab countries of the Middle East; that Ethiopia had provided Israel access to some strategic Red Sea islands like the Dahlack, where Israel had allegedly built military bases to undermine the peace and security of the Arab world; and that the Red Sea should be considered an Arab lake, because 'all' the states surrounding it are Arab" (Kendie 1999, p. 155). Hence Egypt effectively internationalized what had been a domestic problem in Ethiopia, and by doing so it helped Eritrea in its efforts to disengage itself from Ethiopia and, at the same time, took advantage of Ethiopia's deteriorating political, economic, and military situation for its own benefit.

The cold war, combined with ethnic and religious conflict, especially the Arab-Israeli conflict, did have a significant impact on water disputes in the region. In fact, after the United States and the United Kingdom withdrew their financial support for the Aswan High Dam in 1956, Egyptian authorities seized control of the Suez Canal from its British and French owners and subsequently nationalized it, hoping to use the tolls from the canal to finance the dam (Collins 2002, p. 180). Eventually, Egypt would seek assistance for the dam construction from the Soviet Union, effectively providing an opportunity for the West's cold war protagonist to fully insert itself into the politics of the region. Throughout the 1950s and beyond, the cold war, and then the Arab-Israeli conflict, came to have a significant impact on the region's other conflicts, especially over the waters of the Nile River (see, for example, Collins 2002, pp. 180–85; Boutros-Ghali 1997, pp. 321–25).[7]

During the thirty-year uprising for Eritrean independence, Egypt took advantage of a severely weakened Ethiopia to irrigate more lands using water from the Nile River, significantly increasing the amount of land devoted to

7. The British, for example, were furious at the United States for withdrawing financial support for the Aswan High Dam. The British argued that America's action had allowed the Soviets to come to the aid of Egypt and become an important figure in Egyptian political economy for nearly thirty-five years. The withdrawal severely damaged U.S. relations with the Middle East, and the region eventually became an important hot spot for the cold war protagonists. See, for example, Burns (1985).

agriculture. In addition, Egyptian authorities continued to advocate uses for the waters of the Nile River that were considered by the upstream riparian states to be detrimental to their national interests. For example, in 1975 Egyptian authorities announced their intention to begin exploring ways to siphon some of the Nile River's waters for irrigation projects in the Sinai desert, supposedly to help support refugee families from the Gaza Strip and to provide drinking water for pilgrims visiting Jerusalem's holy places (Kendie 1999, p. 156; Kalpakian 2004; Tvedt 2004; Awulachew and others 2012). Such plans by Egypt only reinforced arguments and fears by upstream riparians that downstream states were engaged in water-use processes that ignored and disrespected the interests and values of the upstream countries. The question was, of course, whether Egypt could legally divert the Nile River's waters to projects outside its own boundaries, especially if such diversion were to be undertaken without the express consent of, and at the expense of, all Nile River basin states.

In the 1970s and 1980s, Ethiopia was struck by a tremendous drought, which killed many people and their livestock and created an extraordinary humanitarian problem (Webb, von Braun, and Yohannes 1992; Fradin and Fradin 2008). Faced with such disastrous and deteriorating domestic conditions, Addis Ababa set in motion various policies to reverse the situation and provide an enabling environment for poverty alleviation and human development. One of the first things that the government of Ethiopia did was to initiate a project to determine the feasibility of undertaking irrigation projects based on waters from Lake Tana. As Ethiopian engineers and social scientists began their work on this project, Egyptian authorities reacted swiftly and forcefully. President Anwar el-Sadat warned Ethiopia and its leaders that "any action that would endanger the waters of the Blue Nile will be faced with a firm reaction on the part of Egypt, even if that action should lead to war."[8] While the Ethiopians were suffering tremendously from the 1980s droughts, Egyptians were able to survive and prosper because of the waters that had been stored behind the Aswan High Dam (see, for example, Collins 1997, p. 2).

Yet shortly after Sadat's threat to attack the Ethiopians should the latter attempt to exploit their own resource, the Egyptian president traveled to Israel and formally, and without consulting other Nile riparian states, offered to divert the waters of the Nile River to the patched and thirsty fields of the Negev desert in Israel. In a speech delivered at Haifa, Sadat informed his hosts

8. Quoted in Kendie (1999, p. 157).

of Egypt's intention to transfer some of the water that Egyptians were getting from the Nile River to Israel after the Suez Canal tunnel, which his government was constructing, was completed: "After the tunnel is completed, I'm planning to bring the sweet Nile water—this is the sweetest of the four big rivers of the world—to Sinai. Well, why not send you some of this sweet water to the Negev Desert as good neighbors . . . ? Well, Sinai is on the border with the Negev. Why not? Lots of possibilities, lots of hope."[9]

Kendie (1999, p. 157) and other scholars (Kalpakian 2004, p. 55) have recognized the "ironic contradiction" in Sadat's foreign policy.[10] Sadat, who had claimed to be quite proud of his African heritage (see, for example, Kalpakian 2004, p. 55) was willing to unilaterally divert waters critical to the livelihood of his fellow African neighbors to benefit the inhabitants of Israel. Such an unsolicited offer to the Israelis was in great contrast to the reaction that Sadat had against the Ethiopians, who were simply making plans to use a resource within their territorial borders to deal with domestic problems. In response to Ethiopian protests against the decision by Egypt to divert Nile River waters to the Negev, Sadat proclaimed as follows:

> We do not need permission from Ethiopia or the Soviet Union to divert our Nile water. . . . If Ethiopia takes any action to block our right to the Nile Waters, there will be no alternative for us but to use force. Tampering with the rights of a nation to water is tampering with its life and a decision to go to war on this score is indisputable in the international community. (Quoted in Kalpakian 2004, p. 55)

It is ironic that while Sadat was warning Ethiopia that the latter was interfering with Egypt's rights to the waters of the Nile, Egypt's own actions were preventing Ethiopia from using the waters of the Blue Nile, a river that originates from and runs through Ethiopia, for the latter's domestic needs.

In addition to threatening to wage war against Ethiopia, Egyptian authorities also called on Arab countries to support them in their efforts to

9. W. Claiborne, "Sadat Offers Israel Water from the Nile," *Washington Post*, September 7, 1979, p. A1).

10. Sadat's minister of state for foreign affairs, Boutros Boutros-Ghali, advised Sadat, although unsuccessfully, that the most effective way to secure Egyptian security was for Egypt to improve its relations with Ethiopia and the other upstream riparian states. Hostility toward Ethiopia, Boutros-Ghali argued, would be more detrimental to Egypt than any failure by the Egyptians to irrigate lands in the Negev Desert with precious Nile River waters and secure more friendly relations with Israel. Unfortunately, Boutros-Ghali failed to convince Sadat. See generally Boutros-Ghali (1997).

destabilize the regime in Addis Ababa. Such destabilization was to be undertaken through support for the irredentist movement in what was then the Ethiopian province of Eritrea, whose citizens were involved in a war to secede from Ethiopia and form their own independent polity (Kendie 1999, p. 158; Kalpakian 2004, pp. 55–56), and for supporters of the so-called Greater Somalia movement, which wanted to annex the Ogaden region of Ethiopia and make it part of a Greater Somali nation (see Collins 2002, pp. 214–15).

At the time Sadat was president of Egypt, Mengistu Haile Mariam and the Derg ruled Ethiopia. These were the two strongest political leaders in the Nile River basin. Within their respective spheres of influence, each worked hard to demonize the other. Mengistu created within Ethiopia fears of a Muslim holy war originating in Egypt and sweeping over Ethiopia and across other parts of the region. In addition, Mengistu resurrected long-held fears among Egyptians about the ability of upstream riparian states to block the flow of water, especially from the Blue Nile, into the Nile River. In addition to continuing Egyptian support for Eritrea and the provision of military aid to Somalis in the Ogaden region, Sadat openly referred to Mengistu as a "corrupt Communist puppet of the Soviet Union" and repeated his threat to wage war on Ethiopia should the latter attempt to construct any structures on the Blue Nile that might interfere with the flow of water into the Nile River (quoted in Collins 2002, p. 214).

Boutros Boutros-Ghali, a Coptic Christian and a university professor who would later become secretary-general of the United Nations, advised Sadat to pursue a more pragmatic foreign policy, especially in Egypt's relations with Ethiopia and the upstream riparian states. Boutros-Ghali became Sadat's minister of state for foreign affairs in 1977. As a close adviser to President Sadat on foreign affairs, Boutros-Ghali argued that Cairo should seek to stabilize the Nile River basin region, offer the upstream riparian states, whose economies were less developed, development aid, and cooperate more with Ethiopia, especially in helping the latter secure more domestic stability. He believed strongly that Ethiopia and the other upstream riparian states were not a threat to the security of Egypt and that Egyptian hostility toward these countries would, in the long run, only hurt Egypt and its interests in the region. In fact, he argued, helping the Nile River basin countries, including Ethiopia, develop their domestic economies would benefit Egypt in the long run. The stability that would come about in these countries as a result of the achievement of relatively higher standards of living would be good for Egypt, and the relatively more developed Egyptian economy would secure important markets for its excess output and for the raw materials needed to feed its domestic industries. The prospects for further enhanced trade with the rest of the Nile River basin countries should provide important motivation for Egyptians to cooperate in the development of an international legal regime that is acceptable to all of the watercourse's riparians and which would allow for fair, reasonable, and equitable allocation of the river's waters (Collins 2002, p. 216; Boutros-Ghali 1997).

Unfortunately for Egypt, Sadat declined to accept the recommendations of his minister of state for foreign affairs. Boutros-Ghali later wrote that despite his efforts, Sadat never accepted his views regarding Egyptian attitudes toward Ethiopia and the other upstream riparian states:

> I tried repeatedly to convince Sadat of my views and maintained that Egypt's national interest required us to establish relations with Ethiopia, where 85 percent of the Nile waters originate. To guarantee the flow of the Nile, there is no alternative to cooperation with Ethiopia, particularly in view of the Ethiopian irrigation project at Lake Tana, which could reduce the Nile waters reaching Egypt. As long as relations between Cairo and Addis Ababa were strained or hostile, we risked serious problems. Preserving Nile waters for Egypt was not only an economic and hydrological issue but a question of national survival. As Herodotus declared, "Egypt is the gift of the Nile," and our security depended on the south more than on the east, in spite of Israel's military power (Boutros-Ghali 1997, pp. 321–22).

It was not only Sadat who disagreed with Boutros-Ghali's approach to Nile River governance. Egypt's Islamists also objected to Boutros-Ghali's pragmatism, especially as regards the treatment of Christian Ethiopia and its president, Mengistu Haile Mariam, whom they believed was an infidel. Egypt's Islamists, supported by intellectuals at Cairo's Al-Azhar University, considered the most important center of modern Islamic learning, engaged in a systematic denunciation of what they argued were Ethiopian oppression and exploitation of its Muslim population. Boutros-Ghali relates that when he attempted to remind them that Ethiopia should not be considered the enemy since its highlands were the source of virtually all the waters of the Nile River, he was accused by his fellow citizens of favoring Ethiopia because it was a Christian nation (Boutros-Ghali 1997, p. 63).

Egypt's unilateralists saw the construction of the Aswan High Dam as "the ultimate solution to the tyranny of dependence" on water from the Ethiopian highlands that the country had been subjected to for many years (Collins 2002, p. 217). However, despite the construction of the Aswan dam, Egypt has remained dependent on the waters of the Nile River and its rich Ethiopian nutrients (Collins 2002, p. 218). It appears that Egypt will remain wholly dependent on the Nile River, as it has been for the past 10,000 years. In fact, as many students of the hydrology of the region have discovered, the Nile River is the only meaningful and reliable source of water for Egypt, at least for the foreseeable future. Thus it is important that Egypt improve its relationship with Ethiopia and cooperate with the latter in devising an acceptable allocation formula for the waters of the Nile River.

Mubarak's Pragmatic Nile River Policy

Egypt under Gamal Abdel Nasser and his successor, Anwar el-Sadat, had been extremely unstable. When Hosni Mubarak came to power after the 1981 assasination of Sadat, he sought to stabilize the country and hence was interested in pursuing a more pragmatic policy toward Egypt's neighbors. The architect of Mubarak's policy of rapprochement was Boutros-Ghali. He held fast to his belief that an Egyptian foreign policy that sought "stability in the region through cooperation with upstream riparians" (Collins 2002, p. 216) was the only way to ensure water security for Egypt.

Boutros-Ghali argued that all the upstream Nile River riparian states were extremely poor and politically and militarily incapable of threatening Egyptian security. Nevertheless, these countries were in control of rivers, lakes, and other bodies of water that fed into the Nile River and provided both the precious waters and nutrients needed for Egyptian development. Egypt, he warned, could not ignore these countries and their aspirations. Continued threats against the upstream riparian states, he argued, could only force them to engage in behaviors that could actually endanger Egypt's water security (Collins 2002, p. 216). In his opinion, engaging in dialogue and cooperation with the upstream riparian states, providing economic aid to them, and helping them avail themselves of Egypt's technological expertise, especially in the area of water management, would be more beneficial to Egypt than threats of war (Collins 2002, pp. 216–17).

Mubarak had a more sympathetic ear than Sadat. In 1980 the Organization of African Unity issued what came to be known as the Lagos Plan of Action for Economic Development of Africa, 1980–2000. The report emphasized regional cooperation and the need for Africans to create regional development agencies, which could serve as mechanisms to tackle development issues that were regional in nature. Both Egypt and Sudan used the plea for the development of regional agencies as the foundation on which to build a new policy of regional cooperation, not just in water allocation but in other areas as well.

The new Egyptian policy adopted by the Mubarak government toward the Nile River basin states not only emphasized cooperation but also sought to share Egypt's technological expertise with the relatively less developed upstream riparian states. Egypt's Nile River diplomacy, however, was deliberate and carefully designed not to compromise the country's long-standing position regarding water sharing. This was made clear by the continued insistence by Egyptian authorities that allocation of the waters of the Nile River be governed by the Nile Waters agreements.

Under the leadership of Boutros-Ghali, Egypt's new diplomatic initiative was successful in convincing the Nile River basin states to accept an invitation to attend a conference in Khartoum—the confluence of the White Nile and the Blue Nile—to discuss Nile River governance issues. The conference, which was held in 1983, failed to produce any agreements. However, riparians Egypt, Sudan, Uganda, Zaire (now the Democratic Republic of Congo), and the Central African Republic formed an informal organization called Undugu, which was expected to enhance cooperation across the basin (Collins 2002, p. 224).

Despite these efforts at continued consultation and cooperation by Egyptian leaders, many commentators still argued that they had not seen a major shift in the country's policy toward water allocation issues in the basin. According to Robert O. Collins (1997, p. 6), "The Egyptian policy for Nile Control had not changed in thirty years. There was, in fact, no alternative but continuous dialogue with hesitant and hostile riparians placated by financial support for their local schemes so long as they would not impede the flow of the Nile to Aswan."

Thus despite Mubarak's pragmatic approach to dealing with Nile River basin issues in particular and Egypt's African neighbors in general, Egyptian foreign policy, especially as it related to the waters of the Nile River, did not really see a significant change. In addition to the sponsorship of programs and projects that did not address the critical and twin issues of "Nile control and who owned the Nile" (Collins 1992, p. 7), Egyptian authorities also engaged in other diplomatic efforts designed to maintain the status quo. For example, on July 1, 1993, Mubarak and Meles Zenawi, the president of the transitional government of Ethiopia, signed an agreement in Cairo titled Framework for General Co-operation between the Arab Republic of Egypt and Ethiopia, which reinforced the provisions of the Nile Waters agreements.[11]

11. A copy of the 1993 Zenawi-Mubarak agreement can be found at http://ocid.nacse.org/tfdd/tfdddocs/521ENG.pdf (last visited on June 24, 2013). Zenawi was president of the transitional government of Ethiopia May 28, 1991, to August 22, 1995. In 1995 he was elected prime minister of Ethiopia, a job he held until his death in 2012.

Article 4 of the 1993 Zenawi-Mubarak agreement states as follows: "The two parties agree that the issue of the use of the Nile waters shall be worked out in detail through discussions by experts from both sides, on the basis of the rules and principles of international law." But what was the meaning of "rules and principles of international law"? Did this language imply that the Egyptians could maintain the status quo—the maintenance of the rights granted Egypt and Sudan by the Nile Waters agreements? Articles 5 and 6 of the agreement support this interpretation. Article 5 states that "each party shall refrain from engaging in any activity related to the Nile waters that may cause appreciable harm to the interests of the other party."[12] Under this article, Ethiopia could not undertake any activities on the Blue Nile, the Nile River's most important tributary, without first consulting Egypt, especially if those activities would interfere with the flow of water into the Nile River.

Article 6 of the 1993 Zenawi-Mubarak agreement is even more telling: "The two parties agree on the necessity of the conservation and protection of the Nile waters. In this regard, they undertake to consult and cooperate in projects that are mutually advantageous, such as projects that would enhance the volume of flow and reduce the loss of Nile waters through comprehensive and integrated development schemes."

In other words, Ethiopia could exercise its rights to harvest and use the waters of the Nile River for development purposes, including building structures on the Blue Nile, only after consultations with Egypt. It appears, then, that Ethiopia had effectively subjected itself, through this agreement, to the same constraints as those provided by the Nile Waters agreements.

Some Ethiopian critics of the 1993 Zenawi-Mubarak agreement have claimed that this was a secret deal designed supposedly to reward Mubarak for helping Meles Zenawi capture the apparatus of government in Ethiopia.[13] Although there is no evidence to support such claims, it is important to note that the 1993 Zenawi-Mubarak agreement would effectively allow Egypt to retain its so-called natural and historical rights over the waters of the Nile River, as well as its veto power over construction projects on the Nile River and its tributaries.[14]

12. "Framework for General Cooperation between the Arab Republic of Egypt and Ethiopia," Northwest Alliance for Computational Science and Engineering, Oregon State University, Corvallis, Oregon (http://ocid.nacse.org/tfdd/tfdddocs/521ENG.pdf).

13. See, for example, "Meles-Mubarak 1993 Accord Revisited," *Ethiopian Review*, August 4, 2010 (www.ethiopianreview.com/content/28566).

14. The Zenawi-Mubarak agreement was a bilateral agreement and hence, any rights and obligations created by it would only have legal standing as between the two countries and would have no relevance to the other riparians.

10 *The Grand Ethiopian Renaissance Dam*

The veto power over construction projects on the Nile River and its tributaries granted Egypt by the Anglo-Egyptian treaty of 1929 was reinforced by the bilateral agreement between Egypt and Sudan signed in 1959.[1] The third paragraph of the introduction to the 1959 agreement states that since the 1929 agreement "provided only for the partial use of the Nile waters and did not extend to include a complete control of the River waters," Egypt and Sudan agreed, through the 1959 agreement, to increase their respective shares of the waters of the Nile River. The 1959 agreement also introduced the concept of acquired rights: "The amount of the Nile waters used by the United Arab Republic [Egypt] until this Agreement is signed shall be her acquired right before obtaining the benefits of the Nile Control Projects and the projects which will increase its yield and which projects are referred to in this Agreement."[2]

1. Exchange of Notes between His Majesty's Government in the United Kingdom and the Egyptian Government in Regard to the Use of the Waters of the Nile River for Irrigation Purposes, May 7, 1929, L.N.T.S. 2103; and United Arab Republic and Sudan Agreement (with Annexes) for the Full Utilization of the Nile Waters, Cairo, November 8, 1959, 6519 U.N.T.S. 63. Again, we note that a treaty grants rights only as between those states that are actually bound by the treaty. The states in question here are Egypt and the Republic of Sudan.

2. 1959 bilateral agreement between Egypt and Sudan, art. 1(1). The upstream riparian states, which were not a party to either the 1929 Anglo-Egyptian treaty or the 1959 bilateral agreement between Egypt and Sudan, have refused to recognize the rights granted Egypt and Sudan by these agreements.

Since 1929, Egypt has used both diplomacy and threats of war to discourage the construction of any structures on the Nile River or its tributaries. Thus in April 2011, when Ethiopian authorities in Addis Ababa announced plans to build a hydroelectric dam on the Blue Nile, the Egyptian government immediately launched a series of rhetorical attacks against Ethiopia, including threats of using military force to destroy the dam.[3]

On April 2, 2011, the Ethiopian prime minister, Meles Zenawi, laid the foundation for the Grand Ethiopian Renaissance Dam (GERD).[4] Located in the Benishangul-Gumuz region of Ethiopia and on the Blue Nile River at about forty kilometers (twenty-five miles) east of the Republic of Sudan and owned by the Ethiopian Electric Power Corporation, the dam will create a lake of an estimated 60 billion cubic meters. The Ethiopian government awarded the engineering, procurement, and construction contract, worth US$4.7 billion, to the Italian company Salini Costruttori, and work is expected to be completed by July 2017.[5]

The GERD is part of Ethiopia's efforts to "expand its hydroelectric power capacity" and enhance the country's ability to supply its citizens with energy at affordable prices (Hammond 2013, p. 1). Although Ethiopia's highlands are a rich source of water resources and the country has significant hydroelectric potential, as of 2001 "only 3% of its hydropower potential had been developed" (Hammond 2013, p. 2). By 2013 as many as 83 percent of Ethiopians lacked effective access to electricity. Ethiopia relies on alternative energy sources (for example, biomass fuel), which have significant damaging environmental consequences. The GERD project has the potential to generate as much as 6,000 megawatts of electricity, some of which would be consumed locally in Ethiopia and the rest exported to other countries in the region (Hammond 2013, p. 2).

3. As far as we know, Egyptian authorities have not followed through with such threats. Although international financial institutions have declined to invest in the construction of the Ethiopian dam, there is no evidence to support speculation that Egyptian diplomatic campaigns provided the impetus to the decisions made by these multilateral organizations. It is quite possible that the decision to reject Ethiopia's invitation to participate financially in the construction of the dam was based solely on economic considerations—the latter being the results of a careful cost-benefit analysis.

4. The dam was originally called the Grand Millennium Dam. See "The River Nile: A Dam Nuisance," *The Economist*, April 23, 2011, p. 1.

5. See "Grand Ethiopian Renaissance Dam Project, Benishangul-Gumuz, Ethiopia," WaterTechnology.net (www.water-technology.net/projects/grand-ethiopian-renaissance-dam-africa/).

Despite its abundant water resources, Ethiopia has yet to provide itself with effective and adequate storage facilities that can be used to store excess water collected during the rainy season and used to smooth out the water shortages that occur during the dry season. Hence during the past several decades the country has not been able to effectively manage its droughts and floods. The primary purpose of the GERD, which is expected to stand at 145 meters (475.7 feet) and 1,800 meters (5,906 feet) long with a storage capacity of 63 billion cubic meters of water, however, is the provision of electricity and not enhancement of water management generally (Hammond 2013, p. 2).

Egyptian Reaction to the Grand Ethiopian Renaissance Dam

Shortly after Ethiopia announced that it would start diverting water from the Blue Nile for the purpose of starting construction work on the GERD, the Egyptian government officially launched an effort to confront what appears to be a new dynamism in the upstream riparians' approach to dealing with conflicts affecting use of the waters of the Nile River. Some of these countries seem to be proceeding with development programs, including diverting water from the Nile River or its tributaries, for use in agriculture, commerce, or other economic activities, without adequately consulting Egypt or other riparians.[6]

On May 28, 2013, when Ethiopia began diverting the course of the Blue Nile, Egyptian president Mohamed Morsi asserted that he was not "calling for war" with Ethiopia, but he insisted that "Egypt's water security cannot be violated at all."[7] He added that "all options are open," a statement that was taken in Ethiopia to imply that Egyptians would be willing to wage war against Ethiopia to stop the latter from proceeding with its dam project.[8] Thus in addition to insisting, as previous Egyptian leaders had done, that the upstream riparian states were bound by the Nile Waters agreements and that they must recognize the rights granted to Egypt by these bilateral agreements, Morsi, like his predecessors, also threatened to go to war to defend those rights.

The main concern of Egyptian authorities is "disruption to the river's flow and the detrimental impact this will have on agricultural irrigation, the water-

6. "Egyptian Approaches to the New Developments in the Nile Politics: Water Diplomacy or Water War? Which Way?," Aiga Forum (www.aigaforum.com/articles/Egyptian-Approaches-to-the-Nile-politics.pdf).

7. E. Jobson, "East Africa: Hydropower Politics; the Struggle for Control of the World's Longest River," All Africa, July 1, 2013 (http://allafrica.com/stories/2013070 20664.html), p. 2.

8. Ibid.

way's salinity, its navigability and the country's power generation."[9] In response, Ethiopia points to the results of the study carried out by the Tripartite International Panel of Experts (IPoEs), which, according to Addis Ababa, concludes that the dam will not have any significant damage on water flows to both Egypt and Sudan. That study, however, does not make such definitive conclusions. Ethiopian authorities basically have invoked the Harmon doctrine, which deals with absolute territorial sovereignty, to justify their decision to unilaterally embark on this massive construction project. Under this doctrine, Ethiopia has absolute sovereignty over the water that flows through its territory and can use it in any way it wants regardless of the impact on other riparians (see, for example, McCaffrey 1996, pp. 551–63).

Before he was ousted by the military, Morsi had called a meeting of politicians and other Egyptian governing elites to discuss the results of the International Panel of Experts' study on the impact of the GERD on both Egypt and Sudan.[10] The main objective of the IPoEs' work was "to provide sound review/assessment of the benefits to the three countries and impacts of the GERD to the two downstream countries, Egypt and Sudan."[11] Without the knowledge of the participants, the meeting called by Morsi was transmitted live on Egyptian television.[12] Although most of the rhetoric coming from Egyptian politicians about the GERD was expected, the Ethiopian public was taken aback by the nondiplomatic nature of the comments. Some Egyptian

9. Ibid.

10. The meeting was called the Popular Conference on Egypt's Rights to Nile Water. See Meles Zenawi Memorial, "The International Panel of Experts' Report on the Grand Ethiopian Renaissance Dam," June 15, 2013 (www.meleszenawi.com/the-international-panel-of-experts-report-on-the-grand-ethiopian-renaissance-dam/). This is not the actual report. Rather, the article contains musings about and a "reading" of the report by Ethiopians. The actual report, International Panel of Experts on Grand Ethiopian Renaissance Dam Project (GERDP), "Final Report," Addis Ababa, Ethiopia, May 31, 2013 (www.internationalrivers.org/files/attached-files/international_panel_of_experts_for_ethiopian_renaissance_dam-_final_report_1.pdf) is examined thoroughly later in this chapter.

11. International Panel of Experts, "Final Report."

12. Pakinam el-Sharkawy, President Morsi's adviser for political affairs, released a statement shortly after the broadcast. He explained that, owing to the importance of water security to Egypt, he had decided shortly before the meeting that it would be aired live. Unfortunately, he said, he forgot to inform the attendees that the meeting would be televised live on national television. See, for example, N. el-Behairy, "Morsi Forms Committee on Dam, Meeting Mistakenly Televised," *Cairo Daily News*, June 4, 2013 (www.dailynewsegypt.com/2013/06/04/morsi-forms-committee-on-dam-meeting-mistakenly-televised/).

politicians, unaware that their comments were being transmitted live to a national audience, suggested that the construction of the dam be sabotaged, an activity that, if successfully carried out, would necessarily have been considered by Ethiopians as a declaration of war. Yet another group of participants at the event suggested that the Egyptian government provide military aid to the Oromo Liberation Front in its efforts to destabilize the Ethiopian government and thereby gain either independence or greater autonomy for the region. There was a third suggestion: to spread rumors about false preparations to bomb the dam.[13]

Throughout history, there have been other occasions in which Egyptian authorities have intentionally engaged in "war-mongering rhetoric, support for Ethiopia's enemies and attempts to destabilize Ethiopia."[14] As mentioned earlier, Ethiopians were taken aback by the rhetoric being propagated by Egyptian leaders and public, especially given that over the past several decades, Ethiopians have been making efforts to improve their relations with Egypt and other Nile River riparian states. As Ethiopia's late prime minister Meles Zenawi has argued, the Nile River is the "umbilical cord connecting two great nations," and its management should be based on cooperation, not threats.[15]

All three countries—Egypt, Ethiopia, and Sudan—had agreed that work on the GERD would continue while the Tripartite International Panel of Experts performed its assessment. In addition, Ethiopia was bound by agreement to consult both Egypt and the Sudan before rerouting the waters of the Blue Nile as part of the project. The panel of experts was initiated by Zenawi, who wanted to "demonstrate to Egypt and the Sudan the reality of Ethiopia's commitment to transparency over the GERD."[16]

Morsi, in addition to instructing his ministers of foreign affairs and irrigation and water resources to continue engaging Ethiopians on issues related to the dam, also stated that "Egypt will not give up its right to Nile water" and that "all alternatives [were] being considered." Additionally, he ordered that a national committee consisting of "official, popular and executive authorities" be formed "to deal with the crisis and inform the public about the results." It was reported that during the meeting, while some high-ranking officials suggested that Egypt's military take action against Ethiopia, others argued in favor of "using actors and sports figures to negotiate," and still

13. "Reflections on the Grand Ethiopian Renaissance Dam" (http://somalilandsun.com/index.php/regional/3154-reflections-on-the-grand-ethiopian-renaissance-dam).

14. Meles Zenawi Memorial, "The International Panel of Experts' Report."

15. Ibid.

16. Ibid.

others suggested that Egypt use its diplomatic skills to embarrass Ethiopia internationally.[17]

The failure by Egyptian authorities so far to provide a more nuanced response to the GERD could be because of an increasing awareness in Egypt that it is time to join other riparians in discussions to develop a governance mechanism for the Nile River that reflects the development interests and needs of all the watercourse's states. Even when Morsi was in power, official reaction to the GERD was mixed and often confusing. For example, in May 2013, Mohamed Bahaa el-Din, the Egyptian minister of irrigation and water resources, said that Egypt would not oppose the "establishment of development projects and dams along the Nile River as long as it does not affect water distribution between countries."[18] The minister added that a "tripartite committee involving Egypt, Sudan, and Ethiopia [was] studying the effects that building the [Renaissance] Dam would have" on Egypt and Sudan. Concurring with comments already made by Ethiopian officials concerning the GERD, Egypt's irrigation and water resources minister also stated that "the Renaissance project is a regional project that benefits everyone, including Egypt and Sudan" and that it would not harm either Egypt or Sudan.[19]

Nevertheless, a month later, the Al-Ahram Arabic-language news organization reported that Bahaa el-Din had said that Ethiopia's GERD would be a "disaster" for Egypt, "especially during periods of water scarcity." He added that the government of Egypt "will not give up on one drop of water" and had "started taking procedures that we will not announce," a veiled reference to what many other Egyptian officials, including President Morsi, had declared: war, if necessary, to prevent Ethiopia from completing the dam. Unlike what the minister had said barely a month earlier, he now claimed that "Ethiopia's planned Renaissance Dam project was sure to negatively affect the electricity-generating capacity of Egypt's High Dam."[20]

17. El-Behairy, "Morsi Forms Committee on Dam, Meeting Mistakenly Televised."

18. "Irrigation Minister: Egypt Not Opposed to Renaissance Dam" (www.dailynewsegypt.com/2013/05/12/irrigation-minister-egypt-not-opposed-to-renaissance-dam/).

19. "Egypt Not Opposed to Ethiopia's Renaissance Dam" (www.egypt independent.com/news/egypt-not-opposed-ethiopia-s-renaissance-dam).

20. "Egypt Irrigation Minister Hints at Covert Response to Ethiopia Dam Project" (http://english.ahram.org.eg/NewsContent/1/64/73283/Egypt/Politics-/Egypt-irrigation-minister-hints-at-covert-response.aspx).

Other Views on the Grand Ethiopian Renaissance Dam

In February 2013, the *Sudan Tribune* reported that a senior member of the Saudi cabinet, the deputy defense minister, Prince Khalid bin Sultan, had "unleashed a barrage of attacks against Ethiopia saying that the Horn of Africa nation is posing a threat to the Nile water rights of Egypt and Sudan." In his remarks, the Saudi minister claimed that the Ethiopian dam had the potential to severely damage Sudan in the case of an accident. Speaking to the Arab Water Council in Cairo, Prince Khalid proclaimed that "the Renaissance dam has its capacity of flood waters reaching more than 70 billion cubic meters of water, and is located at an altitude of 700 meters and if it collapsed then Khartoum will drown completely and the impact will even reach the Aswan Dam." He continued: "Egypt is the most affected party from the Ethiopian Renaissance dam because they have no alternative water source compared to other Nile Basin countries and the establishment of the dam 12 kilometers from the Sudanese border is for political plotting rather than for economic gain and constitutes a threat to Egyptian and Sudanese national security."[21] In Khalid's opinion, Ethiopian authorities were determined to harm Arab nations: "There are fingers messing with water resources of Sudan and Egypt which are rooted in the mind and body of Ethiopia. They do not forsake an opportunity to harm Arabs without taking advantage of it."[22]

Prince Khalid went further to state that "the establishment of the dam leads to the transfer of water supply from the front of Lake Nasser to the Ethiopian plateau, which means full Ethiopian control of every drop of water, as well as [causing] an environmental imbalance stirring seismic activity in the region as a result of the massive water weight laden with silt withheld in front of the dam, estimated by experts at more than 63 billion tonnes." The deputy minister concluded his remarks by saying that "the information is alarming and it is important that we do not underestimate the danger at the moment and its repercussions in the future."[23]

Following the hostile and undiplomatic remarks made by Saudi Arabia's deputy defense minister about the GERD, Riyadh acted decisively and quickly

21. "In Unusual Rebuke, Saudi Arabia Accuses Ethiopia of Posing Threats to Sudan and Egypt" (www.sudantribune.com/spip.php?article45666). The dam is actually located on the Blue Nile in the Benishangul-Gumuz region of Ethiopia, about forty kilometers (twenty-five miles) east of the border with the Republic of Sudan.

22. "Saudi Arabia Accuses Ethiopia of Posing Threats to Sudan and Egypt."

23. Ibid. The prince's tirade, however, is not significant, given that it was repudiated, almost immediately, by his country's government. In addition to distancing itself from such remarks, the Government of Saudi Arabia also fired the minister.

and summarily sacked the controversial official. In a statement issued in April 2013, in reaction to Prince Khalid's barrage of attacks on Ethiopia, the government of Saudi Arabia relieved Prince Khalid of his duties, rebuffed his statements, and declared that the minister's remarks "do not reflect the official stance of the Government of the Kingdom of Saudi Arabia." The minister was subsequently replaced, according to Saudi news sources, with Prince Fahd bin Abdullah bin Muhammad.[24]

In 2012 the Sudanese president Omar al-Bashir formally acknowledged the country's support for the Grand Ethiopian Renaissance Dam. He informed Ethiopia's newly appointed ambassador to Sudan, Abadi Zemo, that Sudan was in support of Ethiopia's GERD project.[25] Although Sudan has not yet signed the Nile Basin Initiative's Cooperative Framework Agreement, the endorsement of the GERD by Khartoum is significant. This is so because since 1959, Egypt and Sudan have supported a water allocation regime in the Nile River basin that was designed to ignore the water rights of the upstream riparian states. By implicitly recognizing Ethiopia's right to construct structures on the Blue Nile, Sudan, a downstream riparian, is moving toward the type of reconciliation that could pave the way for greater cooperation in the basin.

The gradual warming of relations between Ethiopia and the Republic of Sudan was further confirmed in 2013. In May that year, the government of Sudan officially rejected statements made by its envoy in Cairo, Kamal Hassan Ali, that the country had opposed the construction of the GERD.[26] The government in Khartoum declared that the Grand Ethiopian Renaissance Dam would not be a threat to Sudan and disclosed that there had been "consultations and understandings" between the three countries—Egypt, Sudan, and Ethiopia—on the GERD project. Khartoum also added that Sudan and Egypt, contrary to speculations in Egypt, would not seek the assistance of the Arab League in dealing with the matter of the GERD. The government of Sudan's spokesperson, Abu Bakr al-Siddiq, also said that his country's ministry of water resources and electricity had assured the government that Ethiopia's dam would not pose any threat to Sudan and that his government is committed to cooperating with both Egypt and Ethiopia on the management of

24. "Saudi King Sacks Minister Who Made Anti-Ethiopia Dam Remarks" (www.sudantribune.com/spip.php?article46307).

25. "Sudan's Bashir Supports Ethiopia's Nile Dam Project" (www.sudantribune.com/spip.php?article41839).

26. "Sudan Downplays Negative Impact of Ethiopian Dam Project" (www.sudantribune.com/spip.php?article46754).

the waters of the Nile River in an effort to agree on a formula for allocation and utilization that is mutually beneficial.[27]

In June 2013 Uganda's president, Yoweri Museveni, announced his country's support for the construction of the Grand Ethiopian Renaissance Dam. In an address to the people of Uganda, Museveni argued that the GERD would spur economic growth not just in Ethiopia but also throughout the region, and that it would help minimize environmental damage caused by the cutting down of trees to produce firewood and the burning of the latter, which produces a lot of pollution.[28] The Ugandan president continued: "It is advisable that chauvinistic statements coming out of Egypt are restrained and through the Nile Valley Organization rational discussions take place." "No African," the president continued, "wants to hurt Egypt; however, Egypt cannot continue to hurt black Africa and the countries of the tropics of Africa."[29] However, while Museveni was supporting Ethiopia's dam-building project on the Blue Nile, Egypt's president, Mohamed Morsi, was warning Ethiopia against continued work on the project and threatening Addis Ababa that he would take whatever measures necessary to prevent Ethiopia from completing the project.[30]

Both the International Monetary Fund and the World Bank have declined to assist Ethiopia in its efforts.[31] However, the government of the People's Republic of China (PRC) has come to the aid of Ethiopia. In spring 2013, Beijing provided Addis Ababa with US$1 billion in loans to build transmission lines linking metropolitan Addis with the GERD.[32] China's commitment to

27. Ibid.

28. "Ethiopia: Uganda Joins Sudan in Support of Grand Ethiopian Renaissance Dam" (http://nazret.com/blog/index.php/2013/06/13/ethiopia-uganda-joins-sudan-in-support-of-grand-ethiopian-renaissance-dam).

29. Ibid.

30. "Uganda Backs Ethiopia in Dam Conflict" (www.dailynewsegypt.com/2013/06/14/uganda-backs-ethiopia-in-dam-conflict/). See also N. el-Behairy, "Morsi: If Our Share of Nile Water Decreases, Our Blood Will Be the Alternative," *Daily News* (Egypt), June 11, 2014 (www.dailynewsegypt.com/2013/06/11/ morsi-if-our-share-of-nile-water-decreases-our-blood-will-be-the-alternative/).

31. Although some commentators have argued that the decision by these multilateral organizations to deny Ethiopia's request for assistance resulted from pressure from Egyptian diplomats, there is no evidence to support that claim. It is most likely the case that the decision to turn down the opportunity to invest in the GERD was based on economic considerations.

32. Meles Zenawi Memorial, "China Lends Ethiopia $1 Billion USD for Mega-Dam Power Lines" (www.meleszenawi.com/china-lends-ethiopia-1-billion-usd-for-mega-dam-power-lines/).

the GERD, of course, is part of Beijing's overall long-term strategy to invest heavily in African infrastructure. Despite the investment made by China, Ethiopia continues to struggle to raise the money to construct the dam.[33]

Meanwhile, Ethiopian authorities have been vocal in their support for the GERD and have made it clear that they will not back down, despite the failure to secure international financing for its construction. For example, while on a tour of various European capitals, the Ethiopian foreign minister, Tedros Adhanom Ghebreyesus, argued that Addis Ababa's approach to the harvesting and utilization of the Nile River waters and the GERD is a "win-win" situation for all riparian states. In Luxembourg, he stated that his government's intention with respect to the waters of the Nile River was to engage in practices that did not impose any significant harm on other riparian states. Hence, he said, Ethiopia will operate "in accordance with the principles of equitable and reasonable utilization."[34] Ethiopia, which contributes most of the water flowing into the Nile River, the minister reiterated, will not accept any water utilization formula that relegates the country to the periphery and effectively makes it a mere bystander. "Ethiopia," Adhanom continued, "would neither accept any proposal that suggested halting the construction of the Dam nor to reduce the Dam's size."[35] Regarding the rerouting of the waters of the Blue Nile on May 28, 2013, the minister stated that this was not a totally new and unique procedure; it was carried out simply to make way for construction work on the GERD to begin.

The Ethiopian foreign minister also indicated that his government believes that Egyptians can benefit significantly from "genuine partnerships" with Ethiopia and the other riparians. Specifically, the minister argued, effective allocation and use of the waters of the Nile River can only be achieved through good faith cooperation between all riparian states.[36]

Adhanom concluded his remarks by noting that throughout the history of the GERD project, Addis Ababa had placed significant emphasis on transparency. The minister stated further that, as part of this effort, the Ethiopian government had established a tripartite committee of experts to examine the proposed GERD and determine its impact (that is, costs and benefits) on all

33. It is expected that construction of the dam would require between $4 billion and $5 billion (U.S.).

34. Tedros Adhanom, "Ethiopia Will Never Halt Construction of the Grand Renaissance Dam," Ethiopiabay.com (http://allafrica.com/stories/201306120099.html).

35. Ibid.

36. Ibid.

relevant parties—specifically, on Ethiopia, Egypt, and Sudan, and generally, on all the other riparian states.[37]

Ethiopia's View of the International Panel of Experts Study

As part of its effort to foster acceptability of the GERD, Addis Ababa established a panel of international experts to study the project and present its results for further action. The panel consisted of ten international experts. Egypt, Ethiopia, and Sudan each had two representatives on the committee; the remaining four were international experts suggested by Ethiopia and accepted by Egypt and Sudan.[38]

Before the report was released to the public, the government of Ethiopia, through various channels, provided summaries and its own interpretations. According to the government of Ethiopia, the Tripartite International Panel of Experts was convened to address the concerns of Egypt and Sudan, helping Egypt and Sudan understand the benefits of the GERD, as well as its costs, and enhancing the ability of the downstream riparians—Egypt and Sudan—to appreciate the "potential shared benefits and impact of the GERD." Specifically, the panel was tasked with "reviewing the design documents of the Grand Ethiopian Renaissance Dam Project; providing transparent information sharing; soliciting understanding of the benefits and costs accruing to each of the three countries; scrutinizing the impacts, if any, of the GERD on the two downstream countries; and building confidence between Ethiopia as an upstream country and the two downstream neighbors." The panel was also expected to propose recommendations to the governments of the three countries "on issues of concern that might be considered in the future."[39]

According to the government of Ethiopia, the panel of experts was expected to perform its job in a transparent manner and make its results accessible to the general public in all three countries. The panel held six meetings and made four visits to the GERD site. Although the study was initiated by Ethiopia, it was supposed to be a cooperative effort. The international team studied the GERD thoroughly. It examined the dam's plans and designs, visited and examined the site, and produced a final report that was submitted to the governments of Egypt, Sudan, and Ethiopia on June 1, 2013.

37. Ibid.
38. Meles Zenawi Memorial, "The International Panel of Experts' Report."
39. Ibid.

In June 2013, Addis Ababa officially released the following comments on the panel's report. First, the GERD project was being carried out "in line with international design criteria and standards."[40] Second, the dam would produce significant benefits for all three countries. Third, the project would not harm or interfere with the flow of water to the two downstream riparian countries. Fourth, it would significantly improve access to affordable energy sources to citizens of the region. Fifth, the new dam would solve the siltation problem in Sudan's and Egypt's dams and significantly enhance water flow. Sixth, the GERD would resolve the problem of seasonal flooding, which was common, especially in Sudan. Seventh, Ethiopia's new dam would significantly reduce loss of water through evaporation and improve the flow of water to, and cut down on the amount of sediment reaching, the Aswan High Dam in Egypt. Eighth, it would provide clean energy for the entire Nile River basin. Finally, the GERD would produce savings in water that would result in more water (for example, from overbank flow and floodplain loss) being available to all riparian states.

Addis Ababa further stated that the report produced two sets of recommendations. One was directed specifically at the government of Ethiopia, and the other collectively at the governments of Ethiopia, Egypt, and Sudan. The recommendations directed specifically at the government of Ethiopia included updating project documents and performing environmental and social impact studies. Addis Ababa said it had accepted the recommendations and had already begun to respond to them. The recommendations to all three countries included conducting "further detailed studies of the water resources and hydrology modeling of the whole Eastern Nile system"; "taking into account that it is proposed to take 5 to 7 years to fill the Dam in order to ensure minimum effects on the flow of the river"; and instructing the three governments to "carry out joint further studies on the environment and social impact and a full trans-boundary environmental impact assessments."[41]

Although Sudan has fully accepted the panel of experts' report, Egypt has not officially clarified its position. Apart from general comments that can be gleaned from the Egyptian press, Cairo has yet to formally engage Sudan and Ethiopia in a conversation about the GERD project. Egypt has been beset by political and economic instability since its recent revolution and the subsequent military overthrow of the elected government of Mohamed Morsi. On June 8, 2014, Abdel Fattah el-Sisi, a former military officer, was installed as the

40. Ibid.
41. Ibid.

country's new president. He promised to "correct the mistakes of the past" and bring peace to the country.[42] There are expectations that the new government, once settled, will make known its position on the Nile River basin and perhaps the GERD. As this monograph goes to press, however, Cairo has not yet presented a clear and well-articulated position on the GERD, nor has it joined its Sudanese and Ethiopian counterparts in good-faith negotiations about the GERD in particular and the management of the Nile River in general.

Final Report of the Panel of Experts

The Tripartite International Panel of Experts was made up of Sherif Mohamady Elsayed and Khaled Hamed of Egypt; Gedion Asfaw, an engineer, and Yilma Seleshi of Ethiopia; Ahmed Eltayeb Ahmed and Deyab Hussein Deyab, an engineer, from the Republic of Sudan; and four international experts: Bernard Yon (environment), John D. M. Roe (socioeconomics), Egon Failer (dam engineering), and Thinus Basson (water resources and hydrological modeling). The number and composition of the panel was agreed to at a meeting of the ministers of water affairs of Egypt, Ethiopia, and Sudan held on November 29, 2011, in Addis Ababa, Ethiopia. At that meeting, the ministers also agreed to the terms of reference and rules of procedure of the panel of experts. The panel's mandate was to "review the design documents of the GERD, provide transparent information sharing and . . . solicit understanding of the benefits and costs accrued to the three countries and impacts if any of the GERD on the two downstream countries so as to build trust and confidence among all parties."[43]

Chapter 1 of the report provides a brief overview of the background to the GERD project, introduces the experts, outlines the timeline for the panel's work, and concludes with a statement about the objectives for the team. A more detailed description of the project is presented in the report's chapter 2. In addition to information on the project's location, this chapter also "summarizes the main characteristics and the technical parameters for the civil, mechanical and electrical works" of the GERD Project (GREDP).[44]

The approach and methodology used by the panel of experts is discussed in chapter 3. The panel reviewed documents about the dam provided by the

42. Y. Basil, "Egypt's New President Vows to 'Correct the Mistakes of the Past,'" *CNN World*, June 9, 2014 (www.cnn.com/2014/06/08/world/africa/egypt-presidential-election/).

43. Introduction to International Panel of Experts, "Final Report."

44. Ibid.

government of Ethiopia, made four visits to the dam site, established a special subcommittee to examine and evaluate geotechnical documents of the GERDP, and undertook field visits to verify information presented in the documents provided them by Addis Ababa. Chapter 4 lists all the documents that were presented to the panel by the government of Ethiopia. Many of these documents were requested by the panel after its first two meetings and after visits to the GERD site.

A summary of the panel's findings and recommendations is presented in chapter 5. These are grouped into three areas: dam safety and engineering, water resources and hydrology, and environment and socioeconomics. With respect to the overall safety of the dam, the panel of experts recommended that verification should be made "under consideration of the additional geological and geotechnical findings. The interaction between the Main Dam and the Powerhouse should be studied and clarified" (5.2.3.3). As part of its work, the panel was tasked with assessing the "downstream impacts of the GERDP on water resources and power generation, both during initial impoundment and under regular operation" (5.3.3.2). The panel determined that operating rules for the existing dam–hydropower installations had not been provided, nor was information offered about the planned operation of the GERDP; that there was no mention of the "planned cascade developments upstream of the GERDP and of the potential impacts of these on the GERDP and further downstream"; that the documents given to the panel did not take into account "losses due to infiltration during first impoundment of the GERD"; and that, given the information available to them, they could not reconcile the "mass balances presented in the report of water between the GERD and the [High Aswan Dam]."[45]

The report states that a "comprehensive study of the GERDP in the context of the Eastern Nile System using a proven, sophisticated and reliable water resource system/hydropower model is strongly recommended to be able to assess and quantify the downstream impacts in detail with confidence" (5.3.3.3). In stating that the GERDP would not harm or interfere with the flow of water to the downstream riparian countries, Ethiopia was, to a certain extent, taking liberties with the report. For, according to the panel, more research would be needed to assess and quantify the impacts of the GERDP on water resources and power generation in Sudan and Egypt (5.3.3.3). The experts were quite cautious in their report and generally recommended that

45. The High Aswan Dam is the largest downstream regulating reservoir.

additional scientific investigations be undertaken to improve confidence in the results obtained.

With respect to environmental and socioeconomic impacts of the GERDP, the panel indicated that the Environmental and Social Impact Assessment (ESIA) Report "satisfies the recommendations of most international funding agencies" (5.4.2.1). The experts cautioned, however, that "the most important water quality issue, which concerns the reduction of dissolved oxygen because of the decay of flooded vegetation and soil, is not adequately addressed in the report" (5.4.2.1). The panel noted that the ESIA report considers the issue of dissolved oxygen demand but concludes that it is insignificant, because Ethiopia only plans to clear out vegetation before undertaking the first water impoundment. In expressing their reservation to this conclusion on the part of the government of Ethiopia, the panel of experts argued that "large amounts of labile carbon in the top soil cannot be removed, . . . vegetation clearance removes only a fraction of the existing biomass, and . . . the area to be cleared is very large (1874 km^2), challenging the technical and financial feasibility of the clearance program" (5.4.2.1). Although Ethiopia was undertaking a study of vegetation clearance, the panel was never presented with its results.

The panel of experts also noted that although the ESIA report provides summaries of the findings of the economic cost–benefit analysis, "which indicated that the GERD is an economically attractive project, . . . the detailed cost/benefit analysis, including downstream costs and benefits was not made available, so the accuracy and reliability of these findings could not be verified by the [panel]" (5.4.2.1). The panel then recommended that to more fully appreciate the impact of the GERDP on the downstream riparian countries, as well as that of dissolved oxygen demand, "the assessment of sediment transport and its organic matter content [be] improved through a sediment monitoring program at the dam site and at least during the rainy season" (5.4.2.2). The panel also recommended that "a trans-boundary impact assessment [be] undertaken for the downstream impact zones within Sudan and Egypt" (5.4.2.2).

After reviewing all the documents about the GERDP provided to them by Ethiopia, which included an initial transboundary environmental impact assessment, and after visiting the dam site, the panel of experts concluded that they did not have enough information to say with certainty what the impact of the GERDP would be on downstream riparian states. Hence they suggested that further scientific investigations be carried out to determine the likely effects of the various aspects of the dam (for example, initial impoundment) on both upstream and downstream states.

11

Governing the Nile River Basin: A Way Forward

At present, the Nile River basin does not have a governance mechanism that is accepted by all the riparian states. The Nile Waters agreements, which are recognized and accepted by the downstream riparians—Egypt and Sudan—but have been rejected by the upstream riparian states, have recognized rights only as between the two parties that are bound by them. The upstream riparians consider the Nile Waters agreements as fostering inequity and unfairness in the allocation of the waters of the Nile River, and they seek to create a new international legal framework for the basin.

However, given that negotiating and producing such an inclusive legal framework is likely to be challenging, it might be necessary for the basin countries to undertake some process through which they can progressively improve on the basin's legal framework to accommodate the interests and concerns of the various riparians. The hope is that through cooperation, they will eventually provide themselves with a fully functioning and inclusive legal framework. Such an approach would ensure that current beneficiaries of the waters of the Nile River are not suddenly subjected to significant reductions in the amount of water available to them. This approach is promising because it appears attractive to countries, such as Egypt, that would be expected to give up some of their current water allocations. These countries would have the opportunity to make necessary adjustments and adapt to the new arrangements. Such adjustments could include adopting water-saving technologies.

The effective legal instrument for governing the Nile River must be one that produces incentives that are comparable with those emanating from existing national and traditional institutional arrangements. Unless there is such compatibility, individuals at the local level are most likely to behave opportunistically and thereby endanger the entire system of transboundary cooperation over the allocation of the Nile River waters. The Nile Waters agreements, for example, allocate all the waters of the Nile River to Egypt and Sudan and leave virtually none to the other riparian states. Such a regulatory framework is incompatible with the development goals and interests of the upstream riparian states and particularly with the communities that lie along the Nile River. It is unrealistic, then, to expect that the citizens or the governments of the upstream riparian states would respect rules that confer enormous costs on them but provide them with virtually no benefits.

On the other hand, designers of any agreement must keep in mind that although Egypt supplies no water to the Nile, the lives of its peoples and, indeed, the country itself are wholly dependent on the waters from the Nile. Hence for any agreement regulating use of the waters of the Nile River to be effective, it must take into consideration Egypt's total dependence on the Nile River for sustenance. In the long run, compliance will depend largely on the willingness of the Nile River basin's relevant stakeholders to accept and respect any compact designed for the governance of the Nile River. How the compact is designed is critical: stakeholders will be more likely to accept and respect a compact that is designed through an inclusive and participatory process. It is only through such a process that the designers can produce an agreement that respects the interests of all riparians, both upstream and downstream.

Although the Nile River riparians may search for guidance in international water management and utilization agreements (such as the UN Watercourses Convention), an effective legal framework for the Nile River basin must be developed by these countries themselves.[1] For such a legal and institutional framework to function effectively and meet the needs of all the states of the Nile River basin, as well as be considered by all riparians as a legal and legitimate water management and utilization mechanism, it must be designed through a democratic process: not only must the governments of all the riparian states participate in the negotiations to produce the governance instrument, but other relevant stakeholders, specifically the various communities that lie along the Nile River and its tributaries, must be provided the facilities to participate fully and effectively in the compacting of the agreement. Thus some form of basinwide consultation must be designed and fully implemented so

1. UN Convention on the Law of the Nonnavigational Uses of International Watercourses, May 21, 1997, G.A. Res. 5/229, U.N. GAOR, 51st sess., 99th plen. mtg., U.N. Doc. A/RES/51/229 (1997).

that all the people who want to participate can do so at a minimum and reasonable cost to themselves. To enhance effective participation, it may be necessary to provide some stakeholders with necessary facilities, which may include language interpreters, telephones, transportation to and from community meeting houses, lodging, food, and access to the Internet.

A report by the World Bank states that "legislators may purposefully base formal law and judicial practice on social norms. In some cases this may consist of simply codifying and modifying existing practices and writing them into law" (World Bank 2002, p. 7). The Nile River basin, of course, is not a homogeneous region, and it is likely to be quite difficult to effectively modify and codify the social norms of all the various stakeholder groups, especially with respect to the allocation and utilization of the waters of the Nile River and its tributaries, and produce a viable and fully functioning formal code of practice. However, bringing together all the relevant stakeholder groups in the basin and engaging them in a democratic design process should enhance their ability to bargain effectively, resolve conflicting interests, and produce laws and institutions that adequately meet the needs of all of the region's constituencies.

Building Viable and Sustainable Multilateral Institutions for Governing the Nile River

To minimize conflict and improve the allocation and sustainability of Nile River basin resources, the multilateral institutions that the Nile River basin countries adopt must be those "that fully and effectively serve the needs of the relevant stakeholders, [and] reflect the latter's interests, beliefs, aspirations, customs and worldview, including such things as social justice and relationship with the environment" (Mbaku 2010, p. 186). The institution that can effectively perform these functions is one that is developed through the good-faith participation and cooperation of all relevant Nile River basin stakeholders and uses relevant time-and-place information provided by the basin's various communities. Such a compact must not only be legal but must also be considered legitimate by all Nile River basin riparians.

In a discussion of legitimacy as it relates to constitutions, Jon Elster (1993, pp. 178–79) argues that the legitimacy of the process through which the constitution is designed, not just the outcome, must be examined and questioned thoroughly. He identifies three types of legitimacy with regard to the process of designing constitutions. The process of designing a constitution is not unlike the process of creating a legal regime for the governance of the Nile River, so Elster's three types of legitimacy are relevant to this discussion. The first is upstream legitimacy, which is concerned with the constituent assembly or committee that is granted the power to produce the constitution. If the constitution-making committee is chosen through a legitimate process (for example, elected by the people through a fair and transparent election), then it will enjoy upstream legitimacy. The second is process legitimacy, which deals with the outcome of the constitution-making process. According to Elster, "If the internal decision-making procedure of the assembly is perceived as undemocratic, the document may be lacking in democratic legitimacy." Elster's third type of legitimacy is downstream legitimacy, which addresses how the final constitutional document is ratified. If the constitution is approved by the people through their representatives, who are elected through free and fair elections, it will have "much stronger claims to embody the popular will"—and hence, will enjoy downstream legitimacy.

An important starting point for the construction of a viable legal and institutional regime for governing the Nile River is a situation analysis. This is a process by which an auditor can determine what types of formal and informal institutions exist, not only in each of the riparian states but also in the various human settlements that dot the banks of the Nile River and its tributaries. Such an audit would allow participants in the institution-building project to determine which functions can be undertaken effectively with existing institutions (both formal and informal) and which require new institutions. It would also allow the riparian states to determine which functions will require management at the regional, national, or community (local) level (Mbaku 2010, p. 187; World Bank 2002, p. 7). This is likely to be a long and expensive process, hence the recommendation that it be undertaken incrementally.

The hydrology of the Nile River basin is likely to change significantly in the long run, owing to climate change and other influences. Hence the legal regime developed must be one that is flexible enough to meet the needs of a changing environment. Such a legal regime can be modeled after a constitution, with emphasis placed on principles rather than specific or codified laws. The courts can, through real cases, interpret these constitutional principles and develop the relevant law.

Consulting stakeholders at the local or community level should allow the designer of institutions in each country to determine what should be undertaken first. For example, assume that the issue to be resolved is to determine the type of water-use rights (which will form part of the overall transboundary water agreement) to grant agrarian farmers located in various communities along the Nile River. Before any decision can be made, there must be full and effective consultation with the farmers whose welfare will be affected by the final protocol. Because policymakers at the center (that is, the capital cities of the various riparian states) are not likely to be in possession of the time-and-place information needed to design a basinwide institutional framework benefiting or servicing the needs of local communities, it is important that these communities are not only consulted but are also provided the facilities to articulate all their concerns and participate fully in the process of institutional design. This calls for all Nile River riparian states to approach the design of laws and institutions for governing water allocation and utilization from a regional perspective: the Nile River and its tributaries should be considered as a regional resource belonging to all the riparian states and governed as such.

Why the emphasis on local consultation? First, elites at the center usually do not have accurate information about economic, social, and political conditions in the local communities. Second, citizens at the local level, especially farmers, have more information than those at the center concerning water demand and supply conditions and the impact of external stimuli such as climate change on local agriculture. Third, rural farmers and other users of water at the local level are in a much better position to determine their social and economic needs and are more likely to have awareness of traditional water management practices that work. Fourth, for institutions to be relevant to the lives of the people of the Nile River basin, these people must be fully involved in their construction. Fifth, institutions designed externally and imposed on the people, as were the colonial-era agreements, will be considered "illegitimate foreign impositions, making compliance difficult and costly" (Mbaku 2010, p. 188). Sixth, as already mentioned, local participation will ensure that the institutions designed are locally focused and hence produce incentives that are compatible with those emanating from informal or traditional institutions. Finally, local participation will allow the people to claim ownership of these institutions and see them as effective instruments for the protection of their rights. Such an acceptance of the institutions should significantly reduce the cost of compliance and allow for more efficient allocation of water resources (Mbaku 2010, p. 188).

Conflict in the Nile River basin must be dealt with at two distinct levels. First, the downstream riparians must accept and deal with the legitimacy of the upstream states' rights to the waters of the Nile River, and the upstream riparians must accept and appreciate the importance of the waters of the Nile River to the lives of the citizens of the downstream countries. Second, all the riparian states must accept and deal effectively with the legitimacy of all internal constituents—that is, the relevant stakeholder groups.

The Role of the Cooperative Framework Agreement

The Nile River Basin Cooperative Framework Agreement (CFA) was designed to provide an inclusive legal mechanism for the equitable and reasonable utilization of the waters of the Nile River, one that was acceptable to all riparians.[2] Unfortunately, the CFA has run into significant problems. First, Egypt and Sudan have formally registered their intention not to sign the agreement. The most important reason for this decision is their disapproval of the CFA's article 14(b), which deals with the issue of water security. Since there was serious dispute over the wording of article 14(b), it was not made part of the CFA's article 14; instead, it was placed in the annex, and the Nile River Basin Commission was charged with resolving the conflict over it. Unfortunately, as this book goes to print, that conflict has not yet been resolved. Although the upstream riparian states recognize the issue of water security, they prefer a version of article 14(b) that does not grant legal standing to the Nile Waters agreements: "Nile Basin States therefore agree, in a spirit of cooperation: . . . (b) not to significantly affect the water security of any other Nile Basin State." The upstream riparian states want the water security mentioned in their version of article 14(b) to be granted by the CFA and not the Nile Waters agreements. Egypt and Sudan, however, want the issue of water security to derive from and be related to the rights that were granted them by the Nile Waters agreements. Hence, they suggested an alternative wording for article 14(b), as follows: "Nile Basin States therefore agree, in a spirit of cooperation: . . . (b) not to adversely affect the water security and *current uses and rights of any other Nile Basin State*" (emphasis added).[3]

As noted earlier, the expression "current uses and rights" put forth by Egypt and Sudan, if accepted by all riparians, would entrench the concept of prior rights, including those created by bilateral treaties that the upstream riparians were not party to and do not accept. The latter have declared that their proposed wording provides for greater levels of fairness and equity in the utilization of the waters of the Nile River.

Although some Nile River riparian states have signed the CFA, the question remains whether this agreement will represent the inclusive legal regime sought by many members of the basin. First, none of the CFA's more than forty articles specifically tackles the critical issue of previous agreements. The fate of the Nile Waters agreements is not specifically addressed by the Cooperative Frame-

2. Agreement on the Nile River Basin Cooperative Framework (www.internationalwaterlaw.org/documents/regionaldocs/Nile_River_Basin_Cooperative_Framework_2010.pdf).

3. Ibid., Annex on Article 14(b).

work Agreement. Second, since Egypt and Sudan have indicated that they do not plan to sign the agreement unless their wording of article 14(b) is accepted by the upstream riparian states, and since the latter have refused to comply with these demands, even if the CFA becomes operative in its present form it would not represent the inclusive legal mechanism desired by all the riparians.

Finally, the most important objective of the CFA was to "balance upper and lower riparian interests" (Ibrahim 2011, p. 304) and end the long-standing conflict between the two parties by providing the basin with an inclusive legal governance mechanism. Unfortunately, the CFA as it currently stands does not resolve that conflict, and without the full participation of both Egypt and Sudan it would not be considered an inclusive legal framework. Thus in its present form it has failed to bring about the necessary balance. Even if, as required by article 42, six countries ratify the agreement and deposit the instruments of ratification with the African Union, and the CFA effectively enters into force, it will still lack the legal legitimacy necessary to resolve the conflict that has dogged issues of water allocation in the Nile River basin since the 1920s. Hence the CFA, without the participation of Egypt and Sudan, remains unilateral action on the part of one party—the upstream riparians—and the conflict will continue. It is, therefore, necessary that all riparian states—both upstream and downstream—engage in efforts to develop an agreement that is acceptable to all of them.

The Way Forward for the Nile River Basin

Obviously, if the upstream riparian states that have signed the Cooperative Framework Agreement—Ethiopia, Kenya, Uganda, Rwanda, Tanzania, and Burundi—successfully ratify the agreement and subsequently bring it into force as prescribed by article 42, the governance crisis in the Nile River basin will remain unresolved since the two downstream riparians—Egypt and Sudan—have refused to sign the agreement. Under these conditions, two legal regimes will exist—one, the Nile Waters agreements of 1929 and 1959, and the other, the Cooperative Framework Agreement, which calls for a new formula for water allocation based on "equity and fairness." Unfortunately, neither regime is acceptable to all the basin's riparians. Three areas of discussion might provide a path forward.

Defining the Problem

Although it is left to all of the Nile River's relevant stakeholders to define the problem that they face, it is clear from previous studies of the political economy of the Nile River basin that the main issue is one of determining or creating a legal mechanism for the equitable and fair allocation of the waters of the Nile River to the bourgeoning populations of the states of the Nile River basin (see, for example, Knobelsdorf 2005–06; Mekonnen 2010; Brunnée and Toope 2002; Ibrahim 2011). Any agreement resulting from negotiations between all the relevant stakeholders should also encourage and "promote beneficial water uses and interstate comity, remove causes of present and future controversies" (Danver 2013, pp. 252–53), enhance poverty alleviation and development efforts in the basin, and provide mechanisms to effectively manage floods and droughts in order to safeguard and protect the property and life of all citizens of the riparian states.

In general, the Nile River should be developed with the primary objective of protecting the rights of the various riparian states and encouraging and enhancing the most economical, fair, and equitable use of water. In maximizing the development of the Nile River and its tributaries, quantitative restraints should be imposed on the extent to which the flow of water can be depleted at specific points on the river, chosen in advance (for example, at Khartoum or Atbara). These restraints should apply to both downstream and upstream states. How such restraints would be imposed and monitored must be determined by the basinwide institution set up for that purpose. In any case, there should be continuous study and reevaluation to enhance the ability of the basin to more efficiently, equitably, and sustainably manage its water resources.

The upstream riparian states have long argued that the Nile Waters agreements provide for a water allocation and utilization scheme that is unfair and inequitable and hence must be discarded. They argue in favor of an allocation formula that fosters equity and fairness. Hence one of the critical issues to be resolved by the riparians as they negotiate a new legal mechanism for governing the Nile River is how to apportion the waters of the Nile River equitably and fairly between all the countries that share this transboundary river system. Perhaps more important, the riparians would have to determine what they mean by "equity" and "reasonableness."

There is need, of course, for the riparian states to jointly take ownership of all the issues associated with the efficient management of the waters of the Nile River. No state should act unilaterally; instead, the approach adopted should be one of cooperation and negotiation in good faith. The concerns of both the upstream and downstream riparians must be thoroughly examined and addressed.

Legal Principles

The Nile River riparian states can benefit from international law regarding the management of transboundary watercourses. Although these countries are free to design and adopt any legal mechanism they desire, it is important to remind them that they are more likely to coexist peacefully and successfully pursue water-utilization projects that enhance human development if the laws and institutions that they adopt for Nile River governance are those that maximize benefits and minimize costs to all the river's relevant stakeholders. Certain legal principles are critical to such a development-oriented legal regime. These include limited territorial sovereignty, "do no harm" to other states, and equitable and reasonable use. These principles are elaborated in chapter 6 of this book. Here, we simply provide an overview and show how their incorporation into a compact between the riparian states would provide them with an effective tool to deal with the management of the Nile River waters and other related issues.

It is generally argued that the concept of territorial sovereignty is critical to the management of a transboundary watercourse (see, for example, Knobelsdorf 2005–06, p. 637). Generally, a sovereign nation has the right "to exert exclusive and unrestricted jurisdiction over the land territory within its boundaries" (Knobelsdorf 2005–06, p. 637). However, when it comes to the exercise of jurisdiction over international watercourses, countries cannot assume absolute sovereignty over their portion of a watercourse as they would with land located entirely in their territorial boundaries, especially in light of modern environmental law (Knobelsdorf 2005–06, p. 638). Under the theory of absolute territorial sovereignty, a country such as Ethiopia can exploit the waters of the Blue Nile without regard to the impact of its actions on downstream riparian states such as Egypt and Sudan. Such an approach is not tenable, especially given that these downstream states depend completely on the waters of the Nile River for their very existence. Hence any effective approach to the management of the waters of the Nile River must adopt limited and not absolute sovereignty as a way to enhance efficient allocation and utilization. Under a limited sovereignty regime, each riparian state's water-use policies must take into consideration the development needs of other riparians. In the case of the Nile River basin, for example, Ethiopia must take into consideration the interests of Egypt and Sudan, and these downstream riparian states must consider Ethiopia's interests and needs.

As stated in section 7(1) of the UN Watercourses Convention, each watercourse state shall "take all appropriate measures to prevent the causing of significant harm to other watercourse States." This maxim is borrowed from the law of property and represents an important part of any legal regime for the management of transboundary water resources. In exploiting and using the portion of a transboundary watercourse within its boundaries, then, a riparian state must make certain that it does not cause harm to other riparians, "either directly through its own actions or by allowing its territory to be used in such a way as to cause injury to another country" (Hall 2004, p. 881). The Nile River riparian states would have to agree on what constitutes harm and whether it is necessary that the harm be substantial before a country is said to have breached its duty of care under the compact.

Although it is the upstream riparian states that have consistently insisted on an approach to water allocation that is based on reasonable and equitable use, it is important that all riparian states, not just the upstream countries, adopt and accept this principle and incorporate it in the compact that they eventually adopt. Both Egypt and Sudan must take into consideration the water needs of the upstream riparian states and agree to cooperate with them in designing and adopting a legal regime that enhances equitable and reasonable allocation and use of the waters of the Nile River and its tributaries. Such an inclusive legal regime should also provide the mechanisms for all the riparian states to deal fully and effectively with issues of basin sustainability, including especially the proper management of the basin's ecosystem.

The Hydrology of the Nile River

Before the negotiations for designing a compact can begin, scientific evaluation of the Nile River and its tributaries should be carried out. There may already exist relevant data on the river and its tributaries, as well as on the Nile River basin; these data can be sourced and additional studies conducted to gather more information.[4] Several documents (for example, the 1929 Anglo-Egyptian treaty) provide data on water levels on the Nile River and these should be consulted. The studies should seek such information as the year with the lowest water level, the year with the highest water level, and the average annual discharge at specific points on the river (for example, Khartoum, Atbara, Aswan High Dam, and Lake Tana). Relatively accurate forecasts of water supply in the Nile River basin should be secured through proper scientific methods. These data would be used to allocate the waters of the Nile River between the upstream and downstream riparian states. Even after this

4. Some of these data can be found in studies of the Nile River and its tributaries carried out by the U.S. Geological Survey.

is done, there must be continuous study and reevaluation to obtain more accurate data to improve management of the river and its resources.

Ethiopia and Egypt have significant expertise in the hydrology of watercourses. Experts should be engaged not only to define the types of data that should be collected but should also help collect those data. John V. Sutcliffe and Yvonne P. Parks (1999, p. 161) provide an effective overview of the hydrological complexity of the Nile River basin system. Effective management of the waters of the Nile River must account for that complexity and be based on a clear understanding of the hydrology of the river and its tributaries and on the political economy of the region. Any activities undertaken by the upstream riparians, including, for example, water diversion schemes and the construction of water storage facilities, can have a significant impact on water availability to the downstream riparians. To minimize conflict, it is necessary to coordinate development in the upstream states with water demand and utilization in the downstream states (Sutcliffe and Parks 1999, p. 161; also see Said 1993). The final compact, then, should also reflect the need for such cooperation and coordination.

Concluding Remarks

The most effective route to a long-term solution to the Nile River dispute is the negotiation by all riparian states of a new inclusive treaty to provide an effective legal mechanism for regulating the utilization of the waters of the river. The challenge is to convince the downstream riparian states, which currently claim rights to the waters of the Nile River that were granted to them through bilateral treaties signed in 1929 and 1959, that it is in their interest to participate fully and effectively in negotiations. There are several reasons why both Egypt and Sudan should cooperate and negotiate in good faith so that the Nile River basin can provide itself with a new and inclusive legal framework.

In the long run, certainty of rights and responsibilities would benefit Egypt and the Republic of Sudan more than the status quo, which is clouded by continuous challenges from the upstream riparians. The upstream riparian states, all of which, except Ethiopia, were colonies during the negotiations that produced the 1929 Anglo-Egyptian treaty, are now independent sovereign nations and have expressly denounced the Nile Waters agreements, asserted their rights to the waters of the Nile River, and indicated that they are not bound by treaties to which they were not contracting parties. Moreover, these countries, notably Ethiopia, have developed significant political, economic, and military power and are now able to withstand Egypt's aggression and threats of war.

Some of these countries have gradually developed the capacity to more effectively undertake construction projects on the Nile River's tributaries that can effectively shut down or at the very least significantly impede the flow of water into the Nile River, especially given that all the Nile tributaries are located in these countries. In fact, Ethiopia is in a position to inflict significant damage on the viability of the Nile River, since as much as 85 percent of the water flowing into the Nile River originates in the Ethiopian highlands.

The global community has changed significantly. In the 1920s, Britain and a few other European countries dominated the political economy of the Nile River basin and hence effectively prevented most of the upstream riparians from asserting their rights to the waters of the Nile River. In fact, Egypt was able to conspire with officials in London to produce a governance regime that allotted all of the waters of the Nile River to Egypt and Sudan. After independence in 1922, Egypt gradually developed into the regional hegemon and came to dominate the political economy of the basin. Today, however, the geopolitics of the Nile River basin are no longer controlled by Britain or, for that matter, even by Egypt, which in recent years has been weakened significantly by domestic issues. More important, the upstream riparian states are now able to assert their interests, and as indicated by Ethiopia's decision to proceed with its Grand Ethiopian Renaissance Dam despite Egyptian opposition, Addis Ababa has also developed the capacity to successfully harvest and utilize the waters of the Nile River and its tributaries.

Finally, given China's decision to lend money to Ethiopia to build transmission lines to link metropolitan Addis Ababa with the Grand Ethiopian Renaissance Dam, and what appears to be Ethiopia's successful campaign to garner financing for the dam from its citizens at home and abroad, a decision by the international donor community to refuse financing for these projects will no longer automatically doom them.[5]

5. "China Lends Ethiopia $1 Billion USD for Mega-Dam Power Lines," Meles Zenawi Memorial (www.meleszenawi.com/china-lends-ethiopia-1-billion-usd-for-mega-dam-power-lines/).

References

Abdalla, I. H. 1971. "The 1959 Nile Waters Agreement in Sudanese-Egyptian Relations." *Middle Eastern Studies* 7, no. 3: 329–41.

Amdetsion, F. 2008. "Scrutinizing the 'Scorpion Problematique': Arguments in Favor of the Continued Relevance of International Law and a Multidisciplinary Approach to Resolving the Nile Dispute." *Texas International Law Journal* 44, nos. 1–2: 1–43.

Awulachew, S. B., and others. 2012. *The Nile River Basin: Water, Agriculture, Governance, and Livelihoods*. Abingdon, U.K.: Routledge.

Baring, E. 2010. *Modern Egypt*. Cambridge University Press.

Bates, R. H. 1981. *Markets and States in Tropical Africa: The Political Basis of Agricultural Policies*. University of California Press.

———. 1983. *Essays on the Political Economy of Rural Africa*. Cambridge University Press.

Beato, A. M. 1994. "Newly Independent and Separating States' Succession to Treaties: Considerations on the Hybrid Dependency of the Republics of the Former Soviet Union." *American University International Law Review* 9, no. 2: 525–58.

Beckert, S. 2004. "Emancipation and Empire: Reconstructing the Worldwide Web of Cotton Production in the Age of the American Civil War." *American Historical Review* 109, no. 5: 1405–38.

Boutros-Ghali, B. 1997. *Egypt's Road to Jerusalem: A Diplomat's Story of the Struggle for Peace in the Middle East*. New York: Random House.

Brace, R. M. 1964. *Morocco, Algeria, Tunisia*. Englewood Cliffs, N.J.: Prentice-Hall.

Brownlie, I. 1990. *Principles of International Law*. Oxford University Press.

Brunnée, J., and S. J. Toope. 2002. "The Changing Nile Basin Regime: Does Law Matter?" *Harvard International Law Journal* 43, no. 1: 105–59.

Bull, H., B. Kingsbury, and A. Roberts, eds. 1992. *Hugo Grotius and International Relations*. London: Clarendon.

Burns, Sir Alan C. 1963. *History of Nigeria*. London: George Allen.

Burns, W. J. 1985. *Economic Aid and American Policy toward Egypt, 1955–81*. State University of New York Press.

Caponera, D. A. 1993. "Legal Aspects of Transboundary River Basins in the Middle East: The Al Asi (Orontes), the Jordan and the Nile." *Natural Resources Journal* 33, no. 3: 629–63.

Carroll, C. M. 1999. "Past and Future Legal Framework of the Nile River Basin." *Georgetown International Environmental Law Review* 12, no. 1: 269–304.

Cliffe, L., and B. Davidson, eds. 1988. *The Long Struggle of Eritrea for Independence and Constructive Peace.* Trenton, N.J.: Red Sea Press.

Cohn, David L. 1956. *The Life and Times of King Cotton.* Oxford University Press.

Collins, R. O. 1997. "The Inscrutable Nile at the Beginning of the New Millennium." Working Paper. Santa Barbara: University of California, Department of History (www.history.ucsb.edu/faculty/Inscrutable%20Nile1.pdf).

———. 2002. *The Nile.* Yale University Press.

Crowder, M. 1987. "Whose Dream Was It Anyway? Twenty-Five Years of African Independence." *African Affairs* 86, no. 342: 7–24.

Degefu, G. T. 2003. *The Nile: Historical, Legal, and Developmental Perspectives; A Warning for the Twenty-First Century*. New York: Trafford.

Dellapenna, J. W. 1994. "Treaties as Instruments for Managing Internationally Shared Water Resources: Restricted Sovereignty vs. Community of Property." *Case Western Reserve Journal of International Law* 26, no. 1: 27–56.

Downie, D. L., K. Brash, and C. Vaughan. 2009. *Climate Change.* Santa Barbara, Calif.: ABC-CLO.

Dunn, J. P. 2005. *Khedive Ismail's Army*. Oxford, U.K.: Routledge.

Earle, E. M. 1926. "Egyptian Cotton and the American Civil War." *Political Science Quarterly* 41, no. 4: 520–45.

Egerton, H. E. 1969. "Colonies and the Mercantile System." In *Imperialism and Colonialism,* edited by G. H. Nadel and P. Curtis, pp. 57–66. New York: Macmillan.

Elhance, A. P. 1999. *Hydropolitics in the Third World: Conflict and Cooperation in International River Basins*. Washington: U.S. Institute of Peace.

Erlich, H. 1994. *Ethiopia and the Middle East.* Boulder, Colo.: Lynne Rienner.

———. 2002. *The Cross and the River: Ethiopia, Egypt and the Nile.* Boulder, Colo.: Lynne Rienner.

Elster, J. 1993. "Constitution-Making in Eastern Europe: Rebuilding the Boat in the Open Sea." *Public Administration* 71, nos. 1–2: 169–217.

Fatton, R., Jr. 1990. "Liberal Democracy in Africa." *Political Science Quarterly* 105, no. 3: 455–73.

Fradin, J., and D. Fradin. 2008. *Droughts: Witness to Disaster*. Washington: National Geographic Society.

Gabre-Selassie, Z. (1975). *Yohannes IV of Ethiopia: A Political Biography*. Oxford, U.K.: Clarendon Press.

Getches, D. H. 2009. *Water Law*. St. Paul, Minn.: Thompson & West.

Godana, B. A. 1985. *Africa's Shared Water Resources.* Boulder, Colo.: Lynne Rienner.

Gopalakrishnan, C., C. Tortajada, and A. K. Biswas, eds. 2005. *Water Institutions: Policies, Performance, and Prospects.* Berlin: Springer.

Guariso, G., and D. Whittington. 1987. "Implications of Ethiopian Water Development for Egypt and Sudan." *International Journal of Water Resources Development* 3, no. 2: 105–14.

Hall, C. G. 1939. *The Origin and Development of Water Rights in South Africa.* Oxford University Press.

Hall, C. G., and H. A. Fagan. 1933. *Water Rights in South Africa* (Cape Town, South Africa: Juta & Company).

Hall, R. E. 2004. "Note: Transboundary Groundwater Management: Opportunities under International Law for Groundwater Management in the United States–Mexico Border Region." *Arizona Journal of International and Comparative Law* 21, no. 3: 873–911.

Hammond, M. 2013. "The Grand Ethiopian Renaissance Dam and the Blue Nile: Implications for Transboundary Water Governance." Discussion Paper 1307. Canberra, Aus.: Global Water Forum" (www.globalwaterforum.org/2013/02/18/the-grand-ethiopian-renaissance-dam-and-the-blue-nile-implications-for-transboundary-water-governance).

Hefny, M., and S. el-Din Amer. 2005. "Egypt and the Nile Basin." *Aquatic Sciences* 67, no. 1: 42–50.

Howell, P. P., and J. A. Allan, eds. 1994. *The Nile: Sharing a Scarce Resource; A Historical and Technical Review of Water Management and of Economical and Legal Issues.* Cambridge University Press.

Ibrahim, A. M. 2011. "The Nile Basin Cooperative Framework Agreement: The Beginning of the End of Egyptian Hydro-Political Hegemony." *Missouri Environmental Law & Policy Review* 18, no. 2: 282–313.

"Impact of Climate Change on the Nile River Basin." 2014. *Rural 21: The International Journal for Rural Development* 48, no. 3: 19–21 (www.rural21.com/english/archive-2005-2011/archive2009-04en/focus/impact-of-climate-change-on-the-nile-river-basin/).

Jacobs, Lisa M. 1993. "Comments: Sharing the Gifts of the Nile: Establishment of a Legal Regime for Nile Waters Management." *Temple International & Comparative Law Journal* 7, no. 1: 95–122.

Jonas, R. 2011. *The Battle of Adwa: African Victory in the Age of Empire.* Harvard University Press.

Kalpakian, J. 2004. *Identity, Conflict, and Cooperation in International River Systems.* Aldershot, U.K.: Ashgate.

Keith, K. J. 1967. "Succession to Bilateral Treaties by Seceding States." *American Journal of International Law* 61, no. 2: 521–46.

Kendie, D. 1999. "Egypt and the Hydro-Politics of the Blue Nile River." *Northeast African Studies* 6, nos. 1–2: 141–69.

Kiplagat, P. K. 1995. "Legal Status of Integration Treaties and the Enforcement of Treaty Obligations: A Look at the COMESA Process." *Denver Journal of International Law and Policy* 23, no. 9: 259–86.

Klare, M. T. 2002. *Resource Wars: The New Landscape of Global Conflict.* New York: Henry Holt.

Kliot, N. 1993. *Water Resources and Conflict in the Middle East.* London: Routledge.

Knobelsdorf, V. 2005–06. "The Nile Waters Agreements: Imposition and Impacts of a Transboundary Legal System." *Columbia Journal of Transnational Law* 44, no. 2: 622–48.

Krishna, R. 1988. "The Legal Regime of the Nile River Basin." In *The Politics of Scarcity: Water in the Middle East,* edited by J. R. Starr and D. C. Stoll, pp. 23–41. Boulder, Colo.: Westview.

Kukk, C. L., and D. A. Deese. 1996–97. "At the Water's Edge: Regional Conflict and Cooperation over Fresh Water." *UCLA Journal of International Law & Foreign Affairs* 21, no. 1: 21–64.

Lautze, J., and M. Giordano. 2005. "Transboundary Water Law in Africa: Development, Nature, and Geography." *Natural Resources Journal* 45, no. 4: 1053–87.

Lester, A. P. 1963. "State Succession to Treaties in the Commonwealth." *International and Comparative Law Quarterly* 12, no. 2: 475–507.

Lie, J. H. S. 2010. *Supporting the Nile Basin Initiative: A Political Analysis "Beyond the River."* Norwegian Agency for Development Cooperation (http://academia.edu/2243972/Supporting_the_Nile_Basin_Initiative_A_Political_Analysis_Beyond_the_River).

Lugard, Sir Frederick. 1926. *The Dual Mandate in British Tropical Africa.* Edinburgh: William Blackwell.

Magubane, B. M. 1979. *The Political Economy of Race and Class in South Africa.* New York: Monthly Review Press.

Makonnen, Yilma. 1983. *International Law and the New States of Africa: A Study of the International Legal Problems of State Succession in the Newly Independent States of East Africa.* Addis Ababa: Ethiopian National Agency for UNESCO.

———. 1986. "State Succession in Africa: Selected Problems." *Collected Courses of The Hague Academy of International Law* 200: 93–148.

Maluwa, T. 1986. "Succession to Treaties and International Fluvial Law in Africa: The Niger Regime." *Netherlands International Law Review* 33, no. 3: 334–70.

———. 1999. *International Law in Post-Colonial Africa.* Dordrecht, Neth.: Kluwer Law International.

Marcus, H. G. 1963. "Ethio-British Negotiations Concerning the Western Border with Sudan, 1896–1902." *Journal of African History* 4, no. 1: 81–94.

Marlowe, J. 1970. *Cromer in Egypt.* London: Elek Books.

Martins, M. A. 1993. "Note: An Alternative Approach to the International Law of State Succession: Lex Naturae and the Dissolution of Yugoslavia." *Syracuse Law Review* 44, no. 3: 1019–58.

Mbaku, J. M. 1997. *Institutions and Reform in Africa: The Public Choice Perspective.* Westport, Conn.: Praeger.

———. 2004. *Institutions and Development in Africa.* Trenton, N.J.: Africa World Press.

———. 2009. "The Public Right to Float through Private Property in Utah: Conatser *v.* Johnson." *Journal of Land, Resources, and Environmental Law* 29, no. 1: 201–25.

———. 2010. *Corruption in Africa: Causes, Consequences, and Cleanups.* Lanham, Md.: Lexington Books.

Mbaku, J. M., and J. O. Ihonvbere, eds. 2003. *The Transition to Democratic Governance in Africa: The Continuing Struggle.* Westport, Conn.: Praeger.

McCaffrey, S. C. 1996. "The Harmon Doctrine One Hundred Years Later: Buried Not Praised." *Natural Resources Journal* 36, no. 3: 549–90.

McCaffrey, S. C., and M. Sinjela. 1998. "The 1997 United Nations Convention on International Watercourses." *American Journal of International Law* 92, no. 1: 97–107.

McCann, J. 1981. "Ethiopia, Britain, and Negotiations for the Lake Tana Dam, 1922–35." *International Journal of African Historical Studies* 14, no. 4: 667–99.

Mekonnen, D. Z. 2010. "The Nile Basin Cooperative Framework Agreement Negotiations and the Adoption of a 'Water Security' Paradigm: Flight into Obscurity or Logical Cul-de-Sac?" *European Journal of International Law* 21, no. 2: 421–40.

———. 2011. "Between the Scylla of Water Security and Charybdis of Benefit Sharing: The Nile Basin Cooperative Framework Agreement—Failed or Just Teetering on the Brink?" *Goettingen Journal of International Law* 3, no. 1: 345–72.

Melesse, A. M. 2011. *Nile River Basin: Hydrology, Climate, and Water Use.* Heidelberg: Springer.

Milkias, P., and G. Metaferia, eds. 2005. *The Battle of Adwa: Reflections on Ethiopia's Historic Victory against European Colonialism.* New York: Algora.

Moret, A. 2001. *The Nile and Egyptian Civilization.* Toronto: General Publishing Company.

Morrison, D. G., R. C. Mitchell, and H. M. Stevenson. 1989. *Black Africa: A Comparative Handbook.* New York: Irvington.

Myers, N. 1989. "Environment and Security." *Foreign Policy*, no. 74 (Spring): 23–41.

Nile Basin Initiative (NBI). 2012. *State of the Nile River Basin 2012* (http://nileis.nile-basin.org/system/files/Nile%20SoB%20Report%20-%20Cover%20Page.pdf)

O'Connell, D. P. 1967. *State Succession in Municipal and International Law*, vol. 1. Cambridge University Press.

Okidi, C. O. 1980. "Legal and Policy Regime of Lake Victoria and Nile Basins." *Indian Journal of International Law* 20, no. 3: 395–447.

———. 1982. "Review of Treaties on Consumptive Utilization of Waters of Lake Victoria and Nile Drainage System." *Natural Resources Journal* 22, no. 1: 161–199.

Organization of African Unity (OAU). 1980. *The Lagos Plan of Action for the Economic Development of Africa, 1980–2000.* Geneva: International Institute for Labor Studies.

Pakenham, T. 1991. *The Scramble for Africa: White Man's Conquest of the Dark Continent from 1876 to 1912.* New York: HarperCollins.

Patel, P. C. 2003. Foreword to *Africa's International Rivers: An Economic Perspective,* by C. W. Sadoff, D. Whittington, and D. Grey, pp. vii–viii. Washington: World Bank (www-wds.worldbank.org/external/default/WDSContentServer/WDSP/IB/2003/02/15/000094946_03020504012641/Rendered/PDF/multi0page.pdf).

Peichert, H. 2000. "The Nile Basin Initiative: A Promising Hydrological Peace Process." In *Cooperation on Transboundary Rivers,* edited by I. al-Baz, V. J. Hartje, and W. Scheumann, pp. 113–32. Baden-Baden, Ger.: Nomos Verlagsgesellschaft.

Plusquellec, H. 1990. "The Gezira Irrigation Scheme in Sudan: Objectives, Design, and Performance." Technical Paper 120. Washington: World Bank (www-wds.worldbank.org/servlet/WDSContentServer/WDSP/IB/1999/12/02/000178830_98101904135320/Rendered/PDF/multi_page.pdf).

Redclif, M. 1986. "Sustainability and the Market: Survival Strategies on the Bolivian Frontier." *Journal of Development Studies* 23, no. 1: 93–105.

Rubenson, S. 1976. *The Survival of Ethiopian Independence.* London: Heinemann.

Rudin, H. R. 1938. *Germans in the Cameroons, 1884–1914: A Case Study in Modern Imperialism.* Yale University Press.

Said, R. 1993. *The River Nile: Geology, Hydrology, and Utilization.* Oxford, U.K.: Pergamon.

Salman, S. M. A. 2012. "The Nile Basin Cooperative Framework Agreement: A Peacefully Unfolding African Spring?" *Water International* 38, no. 1: 17–29.

Schaffer, R. 1981. "Succession to Treaties: South African Practice in the Light of Current Developments in International Law." *International and Comparative Law Quarterly* 30, no. 3: 593–627.

Seaton, E. E., and S. Maliti. 1974. *Tanzania Treaty Practice.* Dar-es-Salaam, Tanzania: Oxford University Press.

Shapland, G. 1997. *Rivers of Discord: International Water Disputes in the Middle East.* New York: St. Martin's Press.

Starr, J. R. 1991. "Water Wars." *Foreign Policy,* no. 82 (Spring): 17–36.

Starr, J. R., and D. C. Stoll, eds. 1988. *The Politics of Scarcity: Water in the Middle East.* Boulder, Colo.: Westview.

Sutcliffe, J. V., and Y. P. Parks. 1999. *The Hydrology of the Nile.* Oxfordshire, U.K.: International Association of Hydrological Sciences.

Suvarna, S. 2006. "Development Aid in an Environmental Context: Using Microfinance to Promote Equitable and Sustainable Water Use in the Nile Basin." *Boston College Environmental Affairs Law Review* 33, no. 2: 449–84.

Swain, A. 2002. "The Nile River Basin Initiative: Too Many Cooks, Too Little Broth." *SAIS Review of International Affairs* 22, no. 2: 293–308.

Theroux, P. 1997. "The Imperiled Nile Delta." *National Geographic* 191, no. 1: 2–35.

El-Tom Hamad, O., and A. el-Battahani. 2005. "Sudan and the Nile Basin." *Aquatic Sciences* 67, no. 1: 28–41.

Tvedt, T. 2004. *The River Nile in the Age of the British: Political Ecology and the Quest for Economic Power.* New York: I. D. Tauris.

Ullendorff, E. 1967. "The Anglo-Ethiopian Treaty of 1902." *Bulletin of the School of Oriental and African Studies* 30, no. 3: 641–54.

United Nations Development Program (UNDP). 2013. *Human Development Report, 2013.* Oxford University Press.

United Nations Environmental Program (UNEP). n.d. *Adapting to Climate Change in the Nile Basin: An Ecosystem-Based Approach to Building Resilience* (www.unep.org/climatechange/adaptation/Portals/133/documents/brochuredesign_final.pdf).

Vivian, C. 2012. *Americans in Egypt, 1770–1915: Explorers, Consuls, Travelers, Soldiers, Missionaries, Writers, and Scientists.* Jefferson, N.C.: McFarland & Company.

Waterbury, J. 1979. *Hydropolitics of the Nile Valley*. Syracuse University Press.

———. 2002. *The Nile Basin: Determinants of Collective Action*. Yale University Press.

Webb, P., J. von Braun, and Y. Yohannes. 1992. *Famine in Ethiopia: Policy Implications of Coping Failure at National and Household Levels.* Washington: International Food Policy Research Institute.

Webster, A. 2006. *The Debate on the Rise of the British Empire.* Manchester University Press.

White, A. S. 1899. *The Expansion of Egypt under Anglo-Egyptian Condominium*. London: Methuen.

Whiteman, M. M. 1964. *Digest of International Law*. U.S. Government Printing Office.

Wiebe, K. 2001. "The Nile River: Potential for Conflict and Cooperation in the Face of Water Degradation." *Natural Resources Journal* 41, no. 3: 731–54.

Wilkinson, H. A. 1975. *The American Doctrine of State Succession*. Westport, Conn.: Greenwood. First published in 1934 by Johns Hopkins University Press.

Wolf, A. T., and others. 1999. "International River Basins of the World." *Water Resources Development* 15, no. 4: 387–427.

World Bank. 2002. *World Development Report, 2002.* Oxford University Press.

———. 2011. *World Development Report, 2011*. Washington.

———. 2012. *World Development Report, 2012*. Oxford University Press.

Wright, P. 1972. *Conflict on the Nile: The Fashoda Incident of 1898.* London: Heinemann, 1972.

Yohannes, O. 2008. *Water Resources and Inter-Riparian Relationships in the Nile Basin: The Search for an Integrative Discourse.* State University of New York Press.

Young, G., and G. Young, Sr. 2002. *Egypt from the Napoleonic Wars down to Cromer and Allenby*. Piscataway, N.J.: Georgias Press.

Zeleke, D. 2005. "Equitable Utilization of Transboundary Watercourses: The Nile Basin and Ethiopia's Rights under International Law." Ph.D. dissertation, University of Vienna.

Index

Lightning Source UK Ltd.
Milton Keynes UK
UKHW010645221222
414324UK00001B/93

9 780815 726555